HISTORY OF TECHNOLOGY

HISTORY OF TECHNOLOGY

History of Technology

Volume Sixteen, 1994

Edited by
Graham Hollister-Short and
Frank A.J.L. James

MANSELL

First published 1994 by
Mansell Publishing Limited, *A Cassell imprint*
Villiers House, 41/47 Strand, London WC2N 5JE, England
387 Park Avenue South, New York, NY 10016-8810, USA

British Library Cataloguing-in-Publication Data

History of technology. — 16th volume (1994)
1. Technology — History — Periodicals
609 T15

ISBN 0–7201–2208–2
ISSN 0307–5451

Library of Congress Catalog
Card Number: 76–648107

Set by Litho Link Ltd, Welshpool, Powys, Wales
Printed and bound in Great Britain by Biddles Ltd, Guildford and King's Lynn

Contents

Editorial

In this volume we are pleased to be able to offer a large collection of papers. Six of them deal with various aspects of the internal history of the discipline. Such weighting is, we believe, correct policy if the history of technology is not to lose proper definition. For otherwise, with apologies to Roy Campbell, one would be all too liable to be left with horse trappings but no horse. In any event, the fact that the economic and social dimensions of the subject are more than adequately dealt with elsewhere means that we are also maintaining a balance in a larger sense.

We would like to draw attention to two of these internal histories in particular, to which we have also given pride of place. Both are concerned with technologies which scarcely existed more than twenty-five years ago. The technologies they describe were needed to develop the oil and gas fields discovered beneath the North Sea, especially those lying beneath the wilder, deeper waters of the high latitudes. William Pike discusses the development of semi-submersible drilling rigs for service in those areas. These are now so well engineered for the extremely hostile environment in which they operate that they are able to stay on station and often continue drilling through all but the most violent storms. Dr Pike recounts a telling incident illustrative of the rigours of the learning process in North Sea conditions to which the rig designers were subjected from the outset. They adopted as a design criterion what they took to be 'once in a hundred years' conditions. The so-called 100-year storm was encountered within weeks by the first rig to start drilling in the southern North Sea. The need to stay on station and continue pumping oil in all conditions gave rise also to the Condeep concept as an alternative to steel jackets. This is Gunnar Nerheim's theme. Here the gigantic scale of the work of fabrication, in concrete, matched the size of the structures which were designed, when in position, to rest on the sea-bed. The structures stand in perhaps 400 feet of water, their upper parts rising a further 100 feet to ensure that the deck is well clear of wave action. Perhaps nothing has characterized Western technology so much as an unrelenting, and unheeding, determination in those directing it to overcome obstacles to their will. North Sea oil engineering exemplifies this as well as revealing the astonishing variety and speed of the engineering response to the problems encountered.

As for the rest of the internal histories, they range in time and place from the late Donald Hill's account of the water clocks of Toledo, built in the last days of Arab rule, before the Christian reconquest of 1085, to the southern United States in the early 1890s. David Lewis's paper on Edward Uehling and his work on the continuous casting process in Birmingham, Alabama, in that decade furnishes a nice reminder that economic motivation may not be, as economic historians usually assume, the sole source of invention. Who shall say that Uehling was not moved as much by compassion for the condition of the manual workers — freed slaves — who worked on the pig-beds? Their lot moved Uehling to remark that had animals been involved in labour of that kind, a society would surely have been formed long before to outlaw such practices.

Of the remaining papers we are especially pleased to be able to include Werner Kroker's presentation on the work and collections of the Deutsches Bergbau Museum at Bochum, as a continuation of a sequence on museums that began in Volume 15 with the Kongsberg Museum in Norway. Angus Buchanan discusses the nature of technological development in what was his inaugural lecture as Professor at the University of Bath. Michael Fores undertakes, with great energy and acumen, we feel, to explore the nature of technological skill. Those who enjoyed reading *Zen and the Art of Motorcycle Maintenance* may find affinities between 'Hamlet without the prince . . .' and Robert Pirsig's description of engineering at the leading edge where there is only the unknown but where also new ideas are given solid form.

Our last comment must be reserved for Melvin Kranzberg's recollections of ICOHTEC and his long association with it as a leading figure in its affairs. The 'International Committee for the History of Technology', for those not familiar with the acronym, was set up in 1968 as a means by which, at the very time of the suppression of the Prague Spring, scholars from Eastern Europe and the West could in some measure continue to meet and thereby insulate as far as possible study of the history of technology from the threat of continual political instability. ICOHTEC has, of course, far transcended that aim in the first 25 years of its existence and has by degrees become the one really authentic international organization concerned with the history of technology.

In Memory of Donald Hill (1922-1994)

I first met Donald Hill in 1974 when he had just completed what I believe will be regarded as his finest undertaking in the history of medieval technology. This was his work in establishing the text, and then translating and critically annotating Ibn al-Razzāz al-Jazarī's *Book of Knowledge and Ingenious Mechanical Devices*, a work written in the northern Mesopotamian city of Diyār Bakr in 1206. Donald was then in full-time employment as an engineer, and nothing better reminds me of the enthusiasm and dedication which he brough to his scholarship than the fact, as he recounted it, that he dealt with the proof-reading with the help of many cups of coffee in a hotel room in Holland in a single weekend, sandwiched between the end of one highly pressurized working week and the beginning of another. Nor was this all. Before Dr Hill retired from work in 1984 he had added a further major translation of a medieval Arabic technological text to stand beside that of al-Jazarī. This was the *Book of Ingenious Devices* written in the 9th century by three brothers known as the Banū Mūsà.

Donald's delight in his explorations of the world of medieval Arabic technology was undimmed to the end. Apart from what I know was the happy task of completing the piece that appears here on the water-clocks of Toledo, he was busy with fresh work. When I last spoke to him, barely two weeks before his death, he was full of enthusiasm for the work he had then just begun on four further medieval Arabic technical manuscripts he had discovered. One of them, which he thought could be dated to 1270, contained a drawing of a pumping device, and it was obvious from his description of it that it matched a later drawing of a similar device to be found in Taccola's *De Machinis* completed in Siena in 1449, raising once again the obscure but important question of the relationship between the two great neighbour cultures, Islam and Christendom, in which he was so greatly interested.

Those who knew Donald will always remember him as an unassuming man of great integrity, who, like all true scholars, was as generous in sharing his knowledge as he was in delighting in the knowledge of others.

Graham Hollister-Short

The Contributors

Professor Hans-Joachim Braun
Universität der Bundeswehr
Hamburg
Holstenhofweg 85
D–2000 Hamburg 70
Germany

Professor R. Angus Buchanan
University of Bath
Claverton Down
Bath BA2 7AY
England

Dr Michael Fores
80 Lexham Gardens
London W8 5JB
England

Dr Donald Hill†
3 Amey Drive
Great Bookham
Surrey KT23 4AL
England

Dr Graham Hollister-Short
History of Science and
Technology Group
Imperial College
London SW7 2AZ
England

Professor Melvin Kranzberg
Georgia Institute of Technology
Atlanta
Georgia 30332–0345
USA

Dr Werner Kroker
Deutsches Bergbau-Museum
am Bergbau-Museum 28
D–4630 Bochum
Germany

Professor W. David Lewis
7030–B Haley Center
Auburn University
Alabama 36849
USA

Dr Gunnar Nerheim
Gamle Harestadv 2A
N–4070 Randaberg
Norway

Dr William J. Pike
c/o Gulf Publishing Company
PO Box 2608 Houston
Texas 77252
USA

Dr Alan Williams
126 Whitehall Road
Woodford Green
Essex IG8 0RZ
England

Notes for Contributors

Contributions are welcome and should be sent to the editors. They are considered on the understanding that they are previously unpublished in English and are not on offer to another journal. Papers in French and German will be considered for publication, but an English summary will be required. The editors will also consider publishing English translations of papers already published in languages other than English. Three copies should be submitted, typed in double spacing (including quotations and notes) with a margin on A4 or American Quarto paper. Include an abstract of 150–200 words and two or three sentences for 'Notes on Contributors'.

It would be appreciated if normal printers' instructions could be used. For example, words to be set in italics should be underlined and *not* put in italics. Authors who have passages originally in Cyrillic or oriental scripts should indicate the system of transliteration they have used. Quotations when long should be inset without quotation marks; when short, in single quotation marks. Spelling should follow the *Oxford English Dictionary*, and arrangement H. Hart, *Rules for Compositors* (Oxford, many editions). Be clear and consistent.

All papers should be rigorously documented, with references to primary and secondary sources typed separately from the text in double spacing and *numbered consecutively*. Cite as follows for books:

> 1. David Gooding, *Experiment and the Making of Meaning: Human Agency in Scientific Observation and Experiment* (Dordrecht, 1990), 54–5.

Subsequent references may be written:

> 3. Gooding, *op. cit.* (1), 43.

Only name the publisher for good reason. For theses, cite University Microfilm order number or at least Dissertations Abstract number. Standard works like DNB, DBB must be thus cited.

And as follows for articles:

> 13. Andrew Nahum, 'The Rotary Aero Engine', *Hist. Tech.*, 1986, 9: 125–66, p. 139.

Line drawings should be drawn boldly in black ink on stout white paper, feint-ruled paper or tracing paper. Photographs should be glossy prints of good contrast and well matched for tonal range. The place of an illustration should be indicated in the margin of the text where it should also be keyed in. Each illustration must be numbered and have a caption. Xerox copies may be sent when the article is first submitted for consideration.

Notes for Contributors

Drilling Technology Transfer between North America and the North Sea: The Semi-Submersible Drilling Unit

WILLIAM J. PIKE

INTRODUCTION

Development of the North Sea oil and gas industry has taken place at a startling pace. When exploration began in 1964, the operating environment of the North Sea was largely unknown. By the late 1980s the industry in the North Sea was generally acknowledged as being at the leading edge of technological advancement. Indeed, by 1985 about 39 per cent of global oil production came from offshore fields, of which about one-quarter was produced in the North Sea.[1] Between the two dates, and especially in the early years, the industry struggled to adapt earlier offshore technology to the demands of the harsher operating environment of the North Sea.

The earliest attempts at adaptation centred on the offshore drilling industry which had largely developed in the much more amicable environment of the Gulf of Mexico. By the advent of exploration in the North Sea, this area had developed, refined and tested offshore drilling technology to a generally sophisticated level. However, equipment brought from the Gulf of Mexico into the North Sea proved almost totally inadequate for drilling operations in more hostile environments. There followed a period of frantic adaptation and redesign of the existing equipment, so much so that by the end of the 1980s North Sea drilling equipment was considered the pinnacle of technology and its designs were being exported world-wide.

This paper seeks to examine the factors which dictated: the need to adapt existing offshore technology, specifically the semi-submersible drilling unit, for use in the North Sea; the nature of the adaptation process and the adaptations; and the general acceptance of North Sea semi-submersible drilling technology as the world-wide standard for the industry. In doing so, it will employ two examples: the development and refinement of the modern semi-submersible drilling structure and the development of motion compensators.

BACKGROUND

The first true offshore well, drilled out of sight of land, was completed off the coast of Louisiana in 1947. Well before that date, however, drilling had taken place over water. Before the turn of the twentieth century wells were being drilled from piers off the coast of California. By the 1920s and 1930s, rigs on barges were exploring the lakes of Venezuela and the swamps and marshes of Louisiana and Texas. By 1964, using experience gained in previous drilling ventures over water, a move into the extended waters of the Gulf of Mexico was well under way.

The simple marine rigs that had began as barges with equipment secured to the top, and which were sunk to rest on the bottom or secured in place by piling, had by the early 1960s developed into the more sophisticated 'jack-up rig'. As the name implies, these rigs had elevating legs attached to the barge. Once over the drilling location, the legs were lowered to the sea-bed and the barge raised above the level of the water. These rigs had two distinct advantages. First, they allowed the rigs to operate in varying, and usually deeper, water depths than the standard barge. Second, during bad weather, the jacking system could raise the barge and isolate it from the interference with operations by heavy seas.

In addition to the jack-up rig, the early 1960s witnessed the widespread acceptance of the drillship. This, as the name indicates, was a ship with a rig mounted topside, which was held over the location by a spread of tensioned anchors. Since the only water-depth limitation with this system was the manageable length of the marine riser system that connected the rig to the well-bore on the sea floor, these rigs offered the potential to drill in much deeper water than the jack-ups, whose operating depth was limited to the length of their legs.[2] The drillship, unlike the jack-up, was not isolated from the action of the waves, however, and often had to suspend operations when weather conditions deteriorated. Both of these rig designs were successfully employed in operations in the Gulf of Mexico, off the coast of Africa and Malaysia, and in other relatively calm waters.[3]

Perhaps the most significant advance in rig design taking place prior to North Sea oil, however, was a new design, the semi-submersible rig (hereinafter referred to as the 'semi', an abbreviation used throughout the industry). The semi was based on the same theory as the drillship in that it would be anchored, under tension, over a drilling location. Like the drillship, it could drill in deeper water than the jack-up. However, the semi had one major advantage over the drillship. When ballasted to drilling draft, it provided a much more stable platform from which to drill — particularly important in the North Sea with its notoriously difficult weather conditions.[4]

With the scramble to begin exploration of the North Sea, all of these rig designs were utilized. The oil companies took whatever rigs were available; there were not many in an industry experiencing a scarcity of drilling equipment. In the United Kingdom Continental Shelf (UKCS), for example, the jack-up *Mr Cap* came from the German sector. An American-built cargo barge, the *Sea Gem*, was modified by the addition of a jack-up system and drilling equipment and shipped to the British sector. The drillship *Glomar IV* sailed under its own power from Galveston, Texas.

Many companies trusted, or at least hoped, that the technology of the Gulf of Mexico would suffice in the North Sea. However, in little more than a year after the initial well was drilled in the UKCS, the first British North Sea oil tragedy was to prove the necessity for rigs specially adapted for the North Sea. In late 1965, much of the lustre was taken from the initial discoveries in the UKCS when the *Sea Gem* toppled and sank after a leg collapsed. The cost was thirteen men out of a crew of thirty-two.

As exploration moved further north, the inadequacies of the equipment became more apparent. Problems often occurred with these rigs because operations were attempted on the basis of experience gained in the calmer waters of the Gulf of Mexico. Wave height was the first problem that had to be conquered. From Gulf of Mexico experience, Howard Brain of Shell Expro was told that the motion of the waves in the North Sea could be compensated for on Shell's new semi, *Staflo*, by simply tying back the brake levers on the anchor winches with a piece of rope. The anchor chains could then feed themselves back and forth under tension to compensate for the heave of the waves. Brain realized the inadequacy of this method when he watched the entire gearbox of an anchor winch rip off and fall into the sea.[5]

Other proposed modifications were the results of overestimation of the requirements for the North Sea. One such idea was the gimballed rotary table.[6] Before experience proved otherwise, it was felt that the strain on the drill pipe created by a rigid rotary table operating under forces of wave motion in the North Sea would lead to drill-string parting and equipment damage. A gimballed rotary table was deemed to be the only solution to the problem, and its development was progressing until experience proved that the existing systems had adequate flexibility for North Sea operations.

Although the problems described above apply to Shell's *Staflo*, every operator experienced similar problems. British Petroleum found, for example, that it had designed a wonderful rig package for its new North Sea semi, *Seaquest*, but had neglected to engineer the hull and column assembly adequately. Experience proved that the three-leg, three-pontoon design was not as seaworthy in the North Sea as had been hoped.[7]

The rig problems were sorted out as early drilling progressed in the North Sea. The first years proved to be important in redefining the climatic conditions in which operations would have to be carried out in the North Sea, and the consequent operational procedures that would have to be followed. Existing weather information had warned the operators of the harsh conditions likely to be encountered.

> Exploration teams had a foretaste of the conditions during early years of exploration in the southern North Sea. Rigs cannot easily break off drilling and run, like ships, for port in a storm. From the beginning, the oil industry expected that, even in shallow southern waters, they would have to be prepared to ride out freak conditions of waves rising to 100 feet as winds gust[ed] to 150 knots on at least one occasion every 100 years. In fact, the first rig to start drilling in the southern North Sea reported that it had encountered the so-called '100 year storm' conditions within a few weeks of moving on to location, so that the deck had to be jacked up a further seven feet to clear the reach of the waves.[8]

Indeed, after five years of operating experience in the North Sea, rig operators Global Marine Inc. calculated that their five drillships working in the area spent between 8.45 and 22 per cent of their operational time in an operations-suspended mode, waiting for the weather to clear.[9] From their experience, they were able to segregate the seasonal phases of weather as follows:

> Summer (5–7 months): predominantly weather in which any type of drilling equipment can be operated effectively.
>
> Spring–autumn (2–4 months): featured by marginal weather during which ships intermittently shut down operations but most semi-submersibles do not.
>
> Winter (2–3 months): that period in which operations of all types are performed relatively inefficiently.[10]

Thus, it became obvious to operators in the North Sea early on that significant adaptations of the traditional offshore drilling technology were required if the area was to be developed. Delays caused by weather, or any other factor, were simply too costly in North Sea operations where, even in the early days, offshore drilling rigs commanded as much as twice the price of their counterparts in other areas of the world.[11]

In response to this challenge, significant refinement and new design was undertaken within the offshore drilling industry in the North Sea, all aimed at minimizing downtime and reducing costs. Of the advances made, perhaps the two most important were the modified semi design and the motion compensator. These two advances characterize the adaptation of the standard semi technology of the 1960s to the North Sea requirements of the 1970s and 1980s.

SEMI-SUBMERSIBLE DESIGN

As discussed earlier, it became apparent in the initial years of North Sea exploration that the drillship was unsuitable for extended operations in the North Sea. While the jack-up was capable of extended operations, its depth limitation, due to leg length, combined with its more dangerous towing configuration and its restricted deck area made it second choice to the semi, which offered a generally suitable compromise between stability, mobility and capacity in most cases.[12]

The first semi for the offshore oil industry, converted from a submersible, sit-on-bottom-type vessel, was the *Bluewater*, which went into operation in the Gulf of Mexico in 1962. In 1964, the first newly constructed semi, *Bluewater II*, went into service, also in the Gulf of Mexico, followed the next year by the *Ocean Driller* and *Ocean Explorer*. 'By 1968, five years after the first unit went into operation, there were 18 units in service and three under construction.' The semis in service represented an investment of approximately $162 million.[13]

The early semis were characterized by many unusual designs. When viewed from above they resembled octagons, pentagons, triangles and, most interestingly, flying wings.[14] Buoyancy was often achieved in the pontoons alone. While this configuration kept the rigs fairly stable by spreading the water plane area, it did not represent the

Figure 1 H-3 semi-submersible platforms. The Aker H-3 semi-submersible platform was designed as a drilling rig for depths up to 300 m (1,000 ft). The first one was ordered in November 1971. The total number built to date is 29. This represents 25 per cent of the world's semi-submersible fleet. Reproduced courtesy of Aker Engineering A/S.

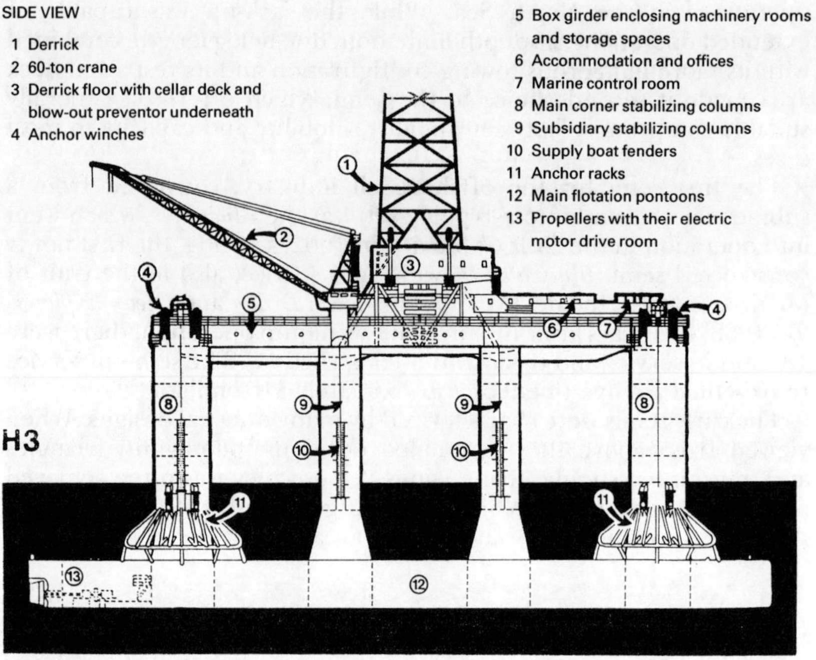

Figure 2 Side view of the Aker H-3 class column-stabilized semi-submersible platform.
Reproduced courtesy of Aker Engineering A/S.

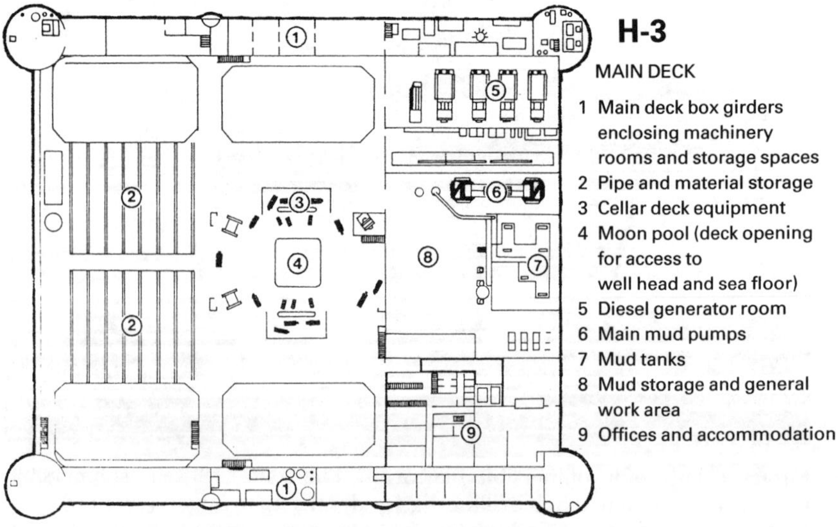

Figure 3 Deck plan view of the Aker H-3.
Reproduced courtesy of Aker Engineering A/S.

optimum in stability.[15] By the mid- to late 1960s a new design was beginning to appear which optimized stability and would be the basis for a general industry-wide standard design.

The new configuration, developed initially for the Gulf of Mexico, featured two parallel pontoons supporting between four and eight massive columns. The columns, for the most part, were hollow voids that provided additional buoyancy.[16] The new design, known as the column-stabilized semi-submersible, allowed greater load capacity, and, more importantly, it greatly enhanced the stability of the vessels, both by increasing buoyant area and by providing a larger area of deck support.

From this initial design was derived the most common type of North Sea semi. Characteristic of the emerging North Sea fleet was the Aker H-3 class, designed by Aker Engineering of Norway.[17] In total, 29 Aker H-3 rigs were delivered to the world's rig fleet between 1972 and 1982, when the design was largely supplanted by the Aker

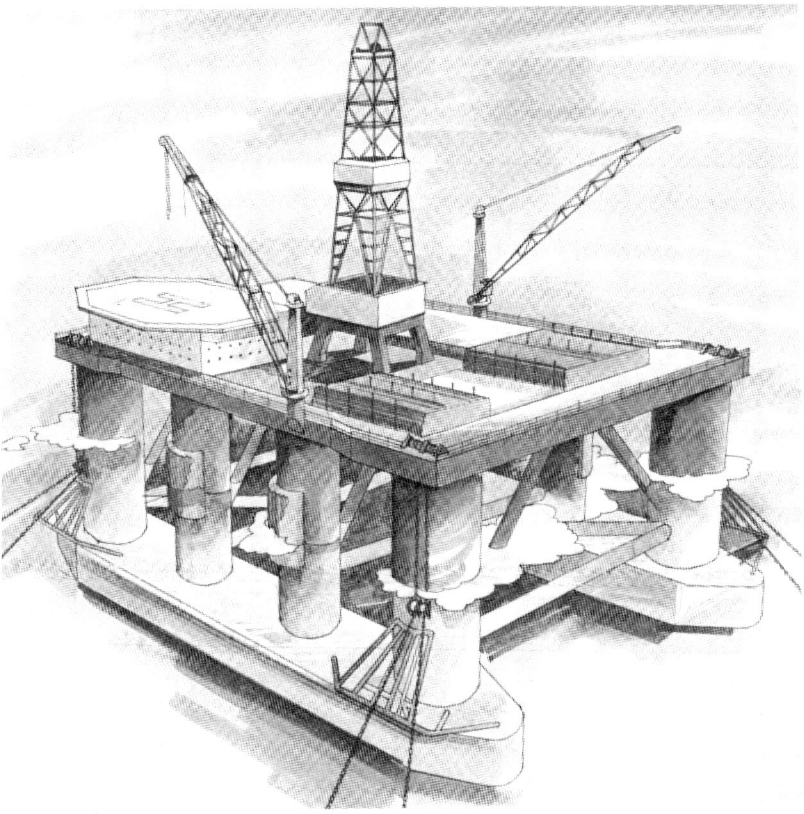

Figure 4 Aker H-3.2E semi-submersible, displacement 26,600 tonnes. Reproduced courtesy of Aker Engineering A/S.

H-3.2 design, an update of the original design.[18] Most of these went to work in the North Sea.

The Aker H-3s, together with similar rigs designed specifically for the North Sea environment, rapidly became the standard for sea-keeping and operational abilities. The rigs of this generation featured:

1. increased deck to water surface clearance to pass the maximum storm waves;
2. additional structural strengthening and redundancy to increase fatigue life and safety;

Figure 5 Aker H4.2 semi-submersible, displacement approx. 40,000 tonnes. Reproduced courtesy of Aker Engineering A/S.

3. addition of propulsion to provide vessel manoeuverability in the event the mooring system [did] not hold in unexpected environments;
4. larger storage capacity to insure sufficient supplies [could] be carried to continue drilling during severe weather periods when supply vessels [could not] approach the rig;
5. addition of special equipment such as motion compensators.[19]

The Aker H-3.2 was an enhancement of the H-3, incorporating additional design improvements. Working to regulations promulgated for the most part by the Norwegian Petroleum Directorate, the rigs were designed to withstand the loss of buoyancy in one leg with a list of no more than 25°. They featured enhanced strength redundancy, improved mooring capabilities, advanced emergency trimming, better motion characteristics and increased load capacity.[20] These qualities allowed them to remain operational through most of the harshest North Sea weather, resulting in considerable savings to drilling programmes for North Sea operators.[21] By 1986, the Aker design had been advanced to the H-4.2 stage. These semis, generally representing the current generation of development, retain the features of the previous generations but are larger in size and provide a much greater variable load rating (Table 1).[22]

Table 1 Specifications of the Aker H-Series semi-submersible.

Class	*H-3*	*H-3.2*	*H-4.2*
Operational water depth (feet)	1,000	1,500	1,968
Drilling depth (feet)	20,000	25,000	25,000
Hull dimensions (feet)	355 × 221	303 × 220	361 × 240
Variable load (mt)	2,900	3,145	4,200

Source: 1991 Directory of Marine Drilling Rigs', *Ocean Industry*, vol. 26, no. 7 (September 1991), pp. 19–118.

Note: Values, especially for variable load, may vary from rig to rig, depending on drilling configuration and modification.

MOTION COMPENSATION

Enhanced station and sea-keeping qualities did not, however, necessarily provide expanded operational limits. While semis might be able to stay on location through a much greater part of the year, it was no guarantee that the drill bit could be kept in contact with the bottom of the hole in a manner capable of allowing drilling to continue. Nor was any surety provided that BOP (blowout prevention) stacks, casing strings and liners could be installed in heavy weather. The development of motion compensation for the drill-string and associated equipment has, therefore, to be seen as a necessary adjunct to the development of more seaworthy vessels.

When drilling from any sort of platform, floating or stationary, the drill bit must be kept in contact with the formation being drilled under constant, controlled pressure. Failure to operate the bit within these parameters results, at best, in poor performance and, at worst, in a damaged or broken bit and ancillary equipment that may require a costly fishing job.[23]

The obvious disadvantage of drilling from a floating platform is that the equipment is rigidly attached to the platform and is subject to the same undulating, reciprocal motion as the platform. The result in early floating operations was that

> either the [drill] bit was spinning freely above bottom while the rig rode a wave crest or it was pounded into the formation when the rig fell into a trough. This problem was solved to a great extent by the use of bumper subs in the drill string. Bumper subs, however, did not provide an altogether satisfactory solution to the critical problems of handling casing strings, BOP stacks, packers, wellheads, test trees, etc.[24]

The result, as mentioned earlier, was either very poor drilling performance or broken equipment; either case cost the operators considerable sums of money.

The solution to the problem was the elimination of a solid connection between the drilling equipment and its supporting vessel, and the drill-string. An elastic connection would absorb the unwanted vertical motion of the vessel.

> Therefore, support [had to be] transmitted through a medium such as a 'spring' with a constant local factor that [would] supply a constant value of support to the load while allowing for vertical motion of the vessel.[25]

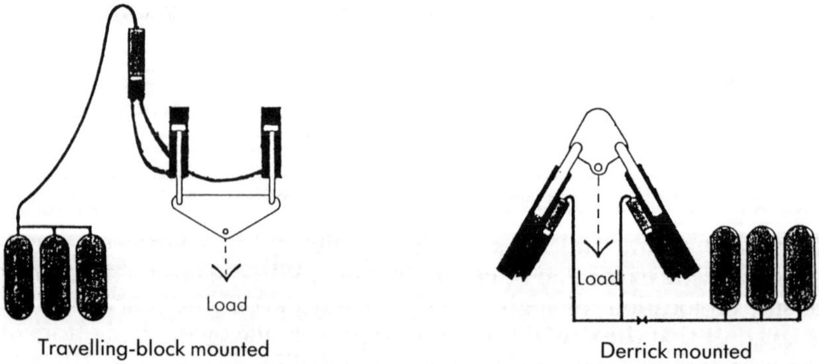

Figure 6 Drill-string compensators.

Source: Maritime Hydraulics

Such a spring was available in the form of contained air. The solution to the problem came in the form of a cylinder, or set of cylinders, in which a piston acted against a large volume of compressed air. It operated on the principle that 'if air is compressed to high pressure in a sufficiently large variable volume system, then small changes in volume of the system will not cause significant changes in the air pressure'.[26] Thus, with the piston acting on the drill-string while the cylinder remained fixed to the vessel, motion compensation in wave heights up to 25 feet was initially achieved.

Although work began on motion compensation systems for drilling in the early 1960s, it was the variable sea states in the North Sea that really gave impetus to their development. The first commercial model appeared in 1970, the product of Vetco Offshore. Within five years, the number of commercial manufacturers had risen to seven.[27]

The primary employment of the first and second generation motion compensators was in the North Sea where continued development, based on the same design parameters, resulted in larger compensation factors. Today, motion compensators are standard equipment on all North Sea semis. Although, like the advanced semi, the motion compensator did not have its origin in the North Sea, its refinement was both crucial to North Sea development and created an enhanced technology, influenced by North Sea operations, that was dispersed throughout the offshore oil industry.

THE EXPORT OF NORTH SEA SEMI-SUBMERSIBLE DESIGN FEATURES

The North Sea has probably forced operators to advance semi drilling technology at a pace much greater than would have been the case otherwise. Operational constrictions of the original designs, combined with the much higher costs of the North Sea, have made the advancement of technology more than a luxury or a pure engineering exercise. The current generation of North Sea semis is able to operate anywhere in the world. They are increasingly being called on to do so as exploration and development move into deeper, less sheltered waters, such as the deeper areas of the Gulf of Mexico and open Atlantic waters off Brazil.

As a mature oil and gas province, plagued by dwindling oil reserves and government-dictated low gas prices, the shallow waters of the Gulf of Mexico are no longer highly attractive to developers. However, much potential is thought, and has been proved in some cases, to exist in deeper waters. These deeper waters will be explored and exploited using primarily North Sea-specified semis which offer more stable, robust and cost-effective operation in deeper waters owing to their refinement for harsh environments.

Similarly, the Atlantic waters off the coast of Brazil are deep, and often violent. North Sea-specified semis have found a market there; in most instances they dominate the market. In all areas where exploration moves off shallow continental shelves and into deep water, it is likely that these vessels will lead the way, not only in drilling, but also in floating production systems in which semis, such as the Aker H-3s, are converted to floating production platforms.

In 1991, out of the 164 semis in the world-wide marine drilling fleet, 131 were of the newer generations which owe a great deal of their design detail to experience gained in North Sea operations.[28] This trend in designing for the harshest environment is likely to continue for the forseeable future.

Notes and References

1. J. Hassan, 'Oil Production Technology, West to East', in D.J. Jeremy, ed., *The Transfer of International Technology* (Aldershot, 1992), 121.

2. In the early development of the semi, marine riser length was limited to around 1,000 feet, primarily because the early semis were small, often without column stabilization, and therefore limited in the deadweight they could support while the riser was in use, and in deck load capacity and deck space when the riser was not in use. Current generation rigs are much larger and current marine riser length, with buoyancy modules fitted, is now limited more by the drilling unit's ability to maintain station over truly deep locations than by any limitations imposed by riser weight or storage requirements; current demonstrated marine riser depth is now at least 10,000 feet. By contrast, operating depth for jack-up rigs will probably not exceed 500–600 feet because optimum jack-up design does not allow for leg weights, or heights when towing, greater than this.

3. Calm is a relative term. The calm conditions described herein refer to the sea state for a great percentage of the year. In the Gulf of Mexico, for instance, sea states normally allow for operation of all types of mobile drilling units about 98 per cent of the year. In the North Sea, that figure is closer to 92 per cent, for modern semis and jack-ups only. In addition, although tropical storms may be much more violent than the most violent North Sea storm, they are more easily predictable and are usually preceded by adequate warning.

4. The semi-submersible drilling unit is formed of multiple hulls (usually called pontoons — there are generally two, but sometimes up to four) from which large columns rise to support a deck structure containing drilling equipment, living quarters and ancillary equipment. They are exceptionally mobile, moving about the world from drilling location to drilling location at towing draft, with their hulls breaking surface as in a conventional vessel. After being positioned on the drilling site, the units are ballasted, submerging their pontoons to depths of thirty or forty feet and isolating them, to a great extent, from the surface action of the waves.

5. Interview with Howard Brain, former Shell drilling engineer and currently Assistant District Manager, Texaco North Sea, Aberdeen, at Aberdeen, 12 May 1989. Shell Expro is a partnership formed between Shell and Exxon for participation in the exploration and development of the United Kingdom Continental Shelf.

6. The rotary table turns the kelly, which turns the drill-pipe with the bit attached at the bottom. It is a large, horizontal, rotating plate on the rig 'floor', generally driven by a chain or gear system from the rotary motor. Four holes in the rotary table receive four pegs of the kelly bushing through which the kelly passes. The kelly is a square or hexagonal forty-foot length of pipe. It

moves freely through the kelly bushing vertically so that the drill-string may be lowered or raised as drilling progresses. Thus, with the kelly bushing engaged in the rotary table, the kelly and the drill-string suspended beneath it have rotary and reciprocal movement.

7. Malcolm Coward, Drilling Manager, British Petroleum Development Ltd, interviewed 18 May 1989, Aberdeen.

8. UK Offshore Operators Association Limited, *The North Sea Achievement* (London, UKOOA, 1989), 10.

9. The five drillships working between 1969 and 1973 were *Glomar III*, *Glomar IV*, *Glomar V*, *Glomar North Sea* and *Glomar Grand Isle*. By the late 1970s, all of these drillships had proved unsuitable for the North Sea and had been withdrawn.

10. A.J. Field, 'Drillship Operations Offshore Northern Europe', paper no. OE–75 206, Offshore Europe Conference, University of Aberdeen, 16–19 September 1975.

11. In the late 1970s, a semi in the Gulf of Mexico commanded $30,000–$40,000 per day. The day rate for a semi in the North Sea during the same period was $65,000–$80,000 per day.

12. With further development, mainly in the area of capacity and operational depth to about 400 feet, the jack-up has played, and continues to play, an important role in North Sea drilling, particularly in shallower coastal and southern waters.

13. D.S. Hammett, 'The Development of Semisubmersibles', paper no. OE–75 204, Offshore Europe Conference, University of Aberdeen, 16–19 September 1975.

14. Examples of these types include: Sedco, Inc.'s *Sedco 135 G* (a triangular design); A/S Norsedrill & Co.'s *Drillmaster* (a pentagon design); Odeco's *Ocean Voyager* (an octagon); and Odeco's *Ocean Driller* and *Ocean Explorer* (flying wings).

15. 'With the spreading of the water plane area, the natural period of the unit increases greatly. The natural period of an SSM [semi-submersible] in heave is normally about 20 seconds, which is far above the everyday wave period experienced during drilling. Reduced motions allow the operator to keep the operation efficiently going during drilling.' Hammett, *op. cit.* (13).

16. The columns did contain some machinery, generally pumping and ballasting equipment.

17. The design for the first H-3 was promoted by Odfjell Drilling and Consulting Company A/S.

18. Other, similar North Sea and hostile water designs appeared, most notably Odeco's 'Ocean Ranger' class and various classes designed by Freide and Goldman of New Orleans. The *Ocean Ranger*, lead vessel of the Odeco class, was lost with all hands off Nova Scotia in a storm when ballast shifting equipment failed to right a serious list.

19. Hammett, *op. cit.* (13).

20. Chuck McCabe, 'First Aker H-3 Begins 8-Year Tour with Statoil', *Ocean Industry*, April 1983, 135–7.

21. Even the most advanced semi designs are forced to suspend operations in truly violent weather. It is unlikely that a mobile drilling unit will ever be developed that can remain operational in the most severe weather. As witness to this fact, a sudden, violent North Sea storm in December 1990 left two semis adrift and one, the *Glomar Arctic III*, severely damaged, although the rig was able to complete operations, before returning for repair. The fact

that no rigs, nor any lives, were lost confirms the safety of the advanced design of the current generation of semis.

22. An example of this class in Sonat's *Arcade Frontier*.

23. When broken equipment, or 'junk', is lost in the hole, it must be 'fished' out. This requires the suspension of drilling activity while special tools are run into the hole to either retrieve (fish) the junk or grind (mill) it into pieces capable of being circulated out of the hole via the drilling mud. Failure to remove a 'fish' usually means loss of the hole, and several million dollars.

24. W.E. Krueger, 'The Development of Motion Compensators', paper no. OE-75 205, Offshore Europe Conference, University of Aberdeen, 16–19 September 1975. The bumper sub is, basically, a tubular expansion joint that acts much like a shock absorber. Since the bumper sub is not of fixed length when in use, it is not suitable for use with the items listed by Krueger as they require very precise measurements of drill-string length to function properly.

25. *Ibid.*

26. *Ibid.*

27. *Ibid.* The manufacturers were: Brown Brothers Rosebank Iron Works; Drilling Systems International; IHC-Holland's Offshore Division; Ocean Science & Engineering, Inc.; Rucker Shaffer/Rucker Control Systems; Vetco Offshore, Inc.; and Western Gear Corporation.

28. Anon., '1991 Directory of Marine Drilling Rigs', *Ocean Industry*, September 1991, 19–118.

The Condeep Concept: The Development and Breakthrough of Concrete Gravity Platforms

GUNNAR NERHEIM

DISCOVERING A NEW OIL AND GAS PROVINCE IN THE NORTH SEA

Following the discovery of the giant Groningen gas field by Shell and Esso in the Netherlands in 1959, the international oil industry turned to the North Sea as an area with very promising geological prospects for finding large accumulations of hydrocarbons, especially gas. The British government awarded the first round of petroleum licences in 1964, followed by the first Norwegian licensing round in 1965. The initial explorations wells on the British shelf were drilled during the 1965 season.[1] Before the end of the year British Petroleum announced that the company had found gas in shallow waters off the south-eastern coast of England. The discovery was evaluated through drilling appraisal wells, and the West Sole Field declared commercial. This discovery was soon followed by others in the same area. Compared to oil industry averages, the first exploration drilling on the British shelf was extremely successful, with six strikes in the first twenty-one wells.[2]

The companies engaged in exploration drilling on the Norwegian shelf had to show much more perseverance. It was not until the end of 1969 that Phillips Petroleum Co., as operator for a group of companies, found the Ekofisk Field. This marked the discovery of the first giant oil field in the North Sea.[3] The finding of reserves capable of supporting large-scale production in Western Europe amazed Europeans and excited even the most cynical members of the oil industry. The Ekofisk discovery triggered frenzied exploration drilling both in the vicinity of Ekofisk and on geological structures further north, in areas with considerably deeper waters. During the first drilling seasons the gales and battering waves of the North Sea proved to the offshore oil industry that the existing American equipment was not well suited to the new environment. The harsher weather conditions to be expected in the most interesting area of all east of the Shetland Islands accelerated the already ongoing technological development to improve the capability of mobile drilling rigs.

Towards the end of 1970, British Petroleum announced the discovery of oil in the Forties Field. In the course of the next few years a string of giant oil and gas fields was discovered on both the British and Norwegian continental shelves, including Frigg (N 1971), Brent (GB 1971), Piper (GB 1973), Ninian (GB 1974) and Statfjord (N 1974). 'The North Sea oil rush was on, and the 1973 oil crisis turned the rush into a roar,' to cite Daniel Yergin.[4] By early 1973 at least 7 billion barrels of recoverable reserves had been found in the British and Norwegian sectors. The North Sea was the richest target laid before the international oil industry since the Middle East discoveries. In the course of three years more recoverable oil was found in the North Sea than in 25 years of operations in the Gulf of Mexico.[5]

THE NORTH SEA ENVIRONMENT AND THE DEVELOPMENT OF NEW TECHNOLOGY

Exploration drilling was conducted by mobile drilling rigs, either jack-ups or semi-submersibles. The jack-up rig is basically a floating barge with elevating legs attached. The drilling rig and all other necessary equipment is placed on the barge. Before drilling the jack-up is towed to the drilling location, where the legs are lowered to the sea-bed and the barge jacked up to an appropriate height above sea-level. The design was introduced in the mid-1950s, while the semi-submersible drilling rig was developed in the early 1960s. This latter consists of multiple hulls, from which large columns rise to support the deck structure. The semi-submersible is a floating vessel. When on location, the pontoons are ballasted down to drilling draft, isolating them to a large extent from the surface action of the waves. The rig is held in place by tensioned anchors. Both the jack-up and the semi-submersible drilling rigs are designed for temporary use. Their main advantage is that they can be moved from one drilling location to the next, if need be from one side of the world to the other.

When offshore petroleum fields are developed for production, however, it is necessary to build *permanent production platforms*, which stand firmly embedded in the sea bottom. Production wells are drilled from the permanent platform, and spread out like the tentacles of an octopus to drain the petroleum reservoir as effectively as possible from one point. This paper focuses on permanent production platforms in the North Sea, and especially on how the idea of a concrete gravity platform, what might be called the Condeep concept, got its breakthrough against many odds. In the late 1970s these platforms came to symbolize the new North Sea technological style.

When North Sea oil exploration gathered momentum, an existing international pool of offshore technology was already there to be tapped. Equipment manufacturers and drilling contractors were engaged in an international trade, following on the heels of the integrated oil companies world-wide. Patents were licensed for international use, and new offshore technological knowledge was published in specialized journals for an international audience. People with offshore skills moved and consulted easily across national boundaries. The transfer of offshore technology was not so much from point to point or place to place as from place to pool to place. The technology of the pool was modified to suit the place.

The style of offshore technology that emerged in the North Sea during the 1970s was shaped to a large degree by the non-technological factors of the cultural context. Thomas P. Hughes defines technological style as 'the technical characteristics that give a machine, process, device or system a distinctive quality. Out of local conditions comes a technology influenced by time and place, a technology with a distinctive style. The local conditions external to the technology can be defined as cultural factors; the technology they shape, a cultural artifact.'[6] Economic, organizational, legislative and entrepreneurial conditions can all be said to fall into the category of cultural factors. In addition the influence of the natural environment came to play a great role in our context. The North Sea weather conditions were a given, but it took engineers and the oil industry some time to adapt, modify and develop a new technology appropriate to that environment. During the first years the oil industry used offshore technologies and practices developed in the Gulf of Mexico, on the assumption that established methods would serve their purpose in any environment. The North Sea soon refuted this assumption. The industry had to invest heavily in the development of new technology, better adapted to the harsh weather conditions in the area.

THE STEEL TRADITION IN OFFSHORE STRUCTURES AND ITS LIMITS

Fixed production platforms in US waters consisted of a series of steel frames joined to form a steel jacket. The completed jacket was skidded onto a special launch barge, transported to the ocean site and launched over a pair of rocker arms. By controlled flooding the jacket was rotated to an upright position and placed on the sea floor. Derrick barges assisted during this operation. The supporting structure was pinned to the sea-bed soil with piles driven through steel sleeves that were part of the structure. On the top of the jacket a steel deck was installed followed by the topsides, the drilling and operating facilities. The drilling rig, oil and gas processing equipment, pumps and compressors, flareboom, landing pad for helicopters, utilities and living quarters for workers were manufactured onshore and packaged into modules not weighing more than the available lift barges had the capacity to handle.

From 1947 until the 1970s the substructure of virtually all offshore oil platforms consisted of steel jackets. The first platforms were supported by a forest of piling because of the small size of steel members being manufactured at the time and the limited lifting capacity and boom length of available floating construction equipment. The first derrick barge for offshore use was commissioned in 1949, with a 150-tonne capacity, followed by a second in 1953, able to lift 250 tonnes. Until 1959, when a platform was installed at a depth of 204 feet, no permanent structures had been installed at water depths beyond 100 feet. When a new generation of derrick barges with lifting capacity of 500 tonnes came into service in 1965, larger and heavier jackets could be designed, conquering depths of more than 300 feet.[7] By 1968 there were 11 derrick barges world-wide with lifting capacity in that range.

It soon became apparent that it was not only the mobile drilling rigs that had to be improved in the North Sea, but the structures of the permanent production platforms as well. When the engineers of BP designed the jackets for the West Sole gas field in 1966, they built on fifteen years of experience with similar structures in the Arabian Gulf at depths up to 26 metres and wave heights of 12 metres. In the course of a few years the North Sea's persistent hammering got the better of these structures. Wave slam, vertical wave-force effects and marine growth all proved to be of greater magnitudes than were expected in the original design specifications.[8] Wave slam on horizontal members in the splash zone, for instance, was greatly underestimated. The wave load proved to be more than three times higher than the design criteria used during construction allowed for. Vertical wave forces and loading due to buoyancy of submerged members were ignored in the original design, with costly repairs as a result.

When it became apparent that several giant oilfields were to be developed in the North Sea, the oil industry suddenly found itself with very strong economic motives to invest in technological innovations in offshore production platforms. The cost of offshore production platforms in the North Sea in the 1970s equalled or exceeded that of nuclear power plants or oil refineries. The oil companies were consequently in a hurry to develop the fields and get the cash flowing to pay for the huge investments.

In the Gulf of Mexico engineers could assume that there would be extended periods of calm and stable weather for fastening the steel jacket to the sea-bed and to install the topsides. In the North Sea major design and construction innovations were needed to adapt the platforms for petroleum reservoirs, which lie under waters where waves measuring 30 metres from crest to trough are common. In the Gulf of Mexico hurricanes could be tracked carefully and offshore crews picked up by helicopter before local conditions became hazardous. That was not easy to accomplish in the North Sea, where as many as 300 workers could be living and working on a single platform with no safe means of departure once a substantial storm developed. Indeed, high winds and seas are so common that all the platforms in the area had to be designed in the expectation that their crews would have to sit tight and ride out the severest storms.

The supporting structure had to secure the topside facilities against the assault of wind, waves, currents and, in some instances, seismic disturbances. The main environmental forces that had to be taken into account were in almost every case related to wave impact. In establishing the design two wave conditions are critical. The first is the maximum wave that might come as a single event during the lifetime of the project. This, the so-called 100-year wave, determines the maximum strength of the structure. The second condition is the cumulative impact that several million waves per year have on the platform. Finally, the supporting structure must have a secure foundation. In soft sea-bottom soils it is at times necessary to drive piles more than 500 feet into the sea-bed before adequate support is achieved.

Should the fixed installations of North Sea fields make use of traditional technological solutions and improve the supporting steel structures, or was new technology called for? In the first case the frontiers of offshore steel technology would have to be pushed to the limit. The oil discoveries in the northern North Sea confronted the engineers with design problems never tackled before anywhere in the world. The discoveries made in the waters east of Shetland were in water depths reaching 500 feet, and where the environment was extremely hostile. The steel tower which was to support the drilling equipment and production facilities for the Brent A platform weighed more than 12,000 tonnes. After the structure had been safely placed on the sea-bed, 32 piles of 72 inches diameter were driven into the soil to a depth of 200 feet, making sure that the completed structure would withstand the imposed loadings.

In the spring of 1973 designers of offshore structures, contractors and operators were wrestling hard with questions concerning the future of platforms in the North Sea. How much would platforms for water depths from 400 feet and over cost? Would extending present designs be the answer, or would radical new approaches and materials have to be introduced? 'The subject is getting so much attention because of the high proportion of total development costs that can be involved in fixed structures', according to *The Oil and Gas Journal*.[9]

In the oil industry lead time is always at a premium. Oil company planners try to keep to a minimum the time between a discovery being proved commercial and production of the first oil or gas. Operations in the North Sea proved to be extremely dependent on the weather, and critical installations had to be put in place within a 'weather window' of short duration. Therefore, structures with short installation periods would be of great interest. A structure which was virtually ready for commissioning on arrival at location would be ideal. This was one reason why several oil companies showed an interest in looking more closely at solutions using reinforced concrete as building material for the substructure of permanent production platforms in the North Sea. But there were other reasons also.

In the early 1970s most European yards with experience from offshore work were already booked for several years, and the traditional shipyards were busy building supertankers. Steel was in short supply and steel prices were rising. Building steel jackets required scarce and highly skilled labour. These factors all contributed to a situation where traditional offshore engineers were willing to consider new technological solutions. French and Norwegian engineers familiar with advanced structural concrete technology felt that using reinforced conrete for supporting structures would be a feasible solution. Concrete lighthouses and military forts had been built in shallow waters, and engineers familiar with the technology strongly believed that the same methods could be used offshore. The size of the offshore structures would be on a different scale, though, as would the water depth and the environmental conditions under which they were to be positioned and operated. Even though reinforced concrete was unknown to the offshore community as a building material, several oil companies seriously began to evaluate design solutions using just that. Development contracts were awarded to encourage designers to come up with solutions suitable for the needs of the industry. The first breakthrough came in the spring of 1971, when Phillips Petroleum chose to build the Ekofisk storage tank in reinforced concrete, using a design by the French engineering firm Doris.

THE FORERUNNER — THE EKOFISK CONCRETE OIL STORAGE TANK

Oil reserves are of little value if they cannot be produced and transported to the market. In the case of the Ekofisk Field, Phillips Petroleum Co. had to solve problems never before faced by the industry. Oil and gas had to be recovered from 70 m of water in a hostile setting, nearly 200 km from shore in the centre of the North Sea. The solutions chosen proved to be both more complex and costly than anticipated, and also contributed to moving the technological frontiers.

The overriding strategy used by Phillips was to develop the field as fast as possible in order to generate some of the cash needed to finance the fixed installations. Within 18 months of the discovery Phillips began producing oil from a temporary production platform at a rate of 42,000 barrels a day.[10] The oil was loaded directly into tankers through pipes attached to nearby buoys. Phillips came to pioneer the use of this system in a deep-water, open-sea situation. The second stage involved the building and installation of permanent production platforms of the traditional steel-jacket type. The development plan would not be completed, however, until pipelines for the transport of oil and gas from Ekofisk to the shore had been built. This would take several years, and in the meantime Phillips would depend on tanker transport, using the offshore loading system. When bad weather made offshore loading impossible, a large storage tank would be needed for intermediate storage. The far less attractive alternative would be to discontinue production.

Phillips asked several firms to develop designs for an offshore storage tank and to bid for the contract. Phillips engineers evaluating the bids were, by training, in favour of a steel tank, but in the end the contract was awarded to the French company Doris, which presented an original solution using reinforced concrete as building material. The French company suggested building the large concrete oil tank in a deep Norwegian fjord, and towing it in a floating position from the fjord to its final location on the Ekofisk Field. By controlled ballasting of the tank it would be lowered slowly onto the sea-bed.[11]

The French company would be in charge of engineering and design, while two experienced Norwegian contracting firms, F. Selmer and Høyer-Ellefsen, founding the joint-venture Norwegian Contractors, would get the subcontract to build the tank. The previous summer, engineers in both of these Norwegian companies had presented ideas using reinforced concrete to build an artificial island with a sheltered harbour at Ekofisk. The solutions were found technically feasible, but costs seemed prohibitive and doubts remained whether an artificial island would be solid enough to withstand the battering of the waves in the area.[12]

The concrete tank for Ekofisk consisted of nine cylindrical storage chambers surrounded by a perforated concrete shell designed to soften the impact of the waves. The total diameter of the structure was 100 m, and it would reach a height of 90 m. It would rest directly on the ocean floor, with 20 m extending above sea-level. One million barrels of oil could be stored.

The foundations of the tank were laid in a dry dock in Stavanger in the autumn of 1971. After some progress had been made, the dock was flooded and the structure towed out to deeper water. Continuous pouring of concrete was very important, and slipforming was used. Concrete was mixed in a floating plant alongside the structure, and to keep a convenient work height the tank was ballasted down in step with the building progress.

The Ekofisk tank was the first concrete storage tank ever built to store oil offshore. The plans required that the tank be completed by the summer of 1972 and towed out before the weather window closed, in this context the time-span between April and October. Owing to delays and new commercial oil discoveries in the Ekofisk area, new concepts for using the tank for additional purposes were developed. The capacity of the oil- and gas-processing equipment originally designed for the Ekofisk Field could not handle the additional quantities from the satellite fields. Phillips engineers decided therefore to install processing equipment on the large area on top of the tank. In addition to holding one million barrels of oil, the tank came to serve as the foundation for the world's largest natural gas processing plant. Another floor was built on top of the tank during the autumn of

1972, and major parts of the equipment were installed inshore before tow-out. In June 1973 six power tugs, with a total capacity of 45,000 hp, towed the Ekofisk tank from Stavanger to its final location (Figure 1).[13] This operation was considered one of the most demanding during the project, and the Norwegian authorities added a piece of drama to the final stages by delaying permission to tow out the tank, demanding that Phillips conduct a new soil investigation of the seabed where the tank was to be grounded at Ekofisk. The towing operation, however, and the ballasting of the tank, proceeded without a hitch.

Figure 1 The Ekofisk concrete storage tank under tow from the building site, Stavanger, Norway in June 1973. One million barrels of oil could be stored in its nine cylindrical cells. This was the first concrete storage tank ever built to hold oil offshore.

Source: Stavanger Aftenblad.

The oil industry followed the Ekofisk tank with great interest through the different stages of construction. It was proof that a concrete structure could be built for offshore purposes and delivered at a price and time schedule competitive with a steel structure. Even before the Ekofisk tank was installed, several other oil companies planning permanent production facilities had invited contractors to tender for both steel and concrete platforms.

THE INVENTION OF THE CONDEEP

French engineers had pioneered the design of the Ekofisk tank, and French engineering firms like C.G. Doris and Sea Tank continued to push concrete technology for offshore oil and gas fields through the design of concrete gravity platforms.[14] They were not alone, however. Engineers employed by Norwegian Contractors were also playing with ideas using concrete for offshore structures and their work, the invention, development and diffusion of the Condeep (Concrete Deepwater Structure), will be followed more closely.

In the spring of 1971 civil engineer Olav Mo at Høyer-Ellefsen thought of designing and building a semi-submersible drilling rig using concrete instead of steel. A large, submerged concrete caisson was to rest on the sea-bed during exploration drilling and serve as a foundation for a traditional steel jacket protruding above sea-level to hold the deck, the derrick and all the usual drilling equipment on top of it. The drilling rig was to be moved from one location to the next by controlled pumping of sea-water in and out of the caisson. Patents were applied for and granted, but further geotechnical studies showed that the idea would be impractical since the sea bottom is often uneven.

The basic design concepts were further developed, however, for permanent production platforms. The caisson was exchanged for cylindrical cells, to be used for oil storage if necessary. Instead of a steel jacket three slender towers in reinforced concrete would be continued above sea-level and out of reach of the 100-year wave. On top of the concrete towers a steel deck containing all the equipment would be installed. Norwegian Contractors saw this kind of concrete gravity substructure as a promising alternative to the use of steel jackets in deep waters. To get a feeling for the feasibility of the basic idea the following crude experiment was executed: seven empty oil barrels were welded together and a tower placed on top; to test the stability of the structure water was let into the oil barrels, and it seemed to behave well both during ballasting and deballasting.[15] Encouraged by the results, Høyer-Ellefsen decided to take development work one step further. The company had comprehensive expertise in the use of concrete, but lacked the necessary mechanical know-how to design and develop a fully fledged production platform. Norwegian Contractors entered into discussions with the Aker Group, by far the most important shipbuilding company in Norway at the time. Aker had recently presented the new Aker H-3 semi-submersible mobile drilling rig and signed the first building contracts. It was decided to form a joint venture between Norwegian Contractors and the Aker Group, to be called the Condeep Group, to further develop and market the Condeep. Aker undertook the design systems for ballasting, crude oil handling and the steel deck on top of the concrete towers. Before the end of 1971 the Norwegian partners could present the oil industry with the first prototype of the Condeep.

To get oil companies bound by tradition to grasp the concepts behind this product innovation was no easy task in itself, but getting them to award the innovators a building contract in the range of £20 million seemed almost impossible. It took time, for instance, to convince American petroleum engineers that concrete was a lasting material in sea-water. It helped that Norwegian Contractors could point to the hundreds of concrete jetties built by the group in Norway, several of them more than 50 years old and still in daily use. It was a great encouragement when the Condeep Group was awarded a development contract by BP in the summer of 1972 to design a Condeep gravity platform for the Forties Field. This contract was worth several million kroner and enabled the Norwegian companies to cover their R&D expenses. In the final bid evalution BP and its consultants concluded that the Norwegian design was technologically advanced and represented a viable solution for a concrete platform on the Forties Field. In the end, however, BP chose to build a huge steel platform.

Nevertheless, Norwegian Contractors experienced the development contract with BP as a breakthrough. One of the major oil companies had shown serious interest, and had been willing to back up its interest by spending a substantial sum of money on the further development of the Condeep concept.

During the autumn of 1972 the Condeep Group also tendered for a contract to build a concrete production platform for the Frigg Field, where the French oil company Elf was operator. By February 1973 it was evident that Elf was in favour of a steel jacket platform, to be built by the French UIE-Group. The price of the Norwegian Condeep and the French Sea Tank concrete designs were approximately the same, although 15 per cent higher than the steel jacket platform.[16]

The difference in price between concrete and steel was not large, considering the arguments used by the Condeep Group. The Condeep could be completed inshore. The deck would be mated to the substructure and equipment installed prior to tow-out, resulting in considerable reduction in costly offshore work. Further, the platform did not need any piling offshore after installation. The greatly increased deck area and the larger load that could be accommodated on a concrete gravity structure, compared to the traditional ones, permitted increased production capacity on each platform, and this in turn might reduce the total number of platforms required to develop a field. Installation of concrete substructures would lead to reduced maintenance costs. The experience in the North Sea so far had shown that maintenance of steel jackets was very expensive, while high-quality concrete would require very modest expenditures. An in-house study made by Norwegian Contractors concluded that concrete platforms could cost up to 25 per cent more than a steel jacket, and still have the same total economy. Arguments like these, however, did not carry great weight with the French oil companies. They decided to build a traditional steel jacket platform, and were happy to place the contract with a French yard.

During the spring of 1973 the Condeep Group tendered for contracts with Mobil for a production platform on the Beryl Field, and with Shell for the second platform on the Brent Field — both on the British shelf. The fields lay in deep water, and the Condeep Group hoped that this would weigh in favour of concrete. If Norwegian Contractors was awarded one or both of these contracts, construction would commence immediately at the dry dock in Stavanger, since both platforms would have to be completed and ready for tow-out in the summer of 1975.

BUILDING THE FIRST CONDEEPS

Friday the 13th of July 1973 brought luck to the Condeep Group. Mobil North Sea ordered the first Condeep ever to be built, the Beryl A, followed by a new order four weeks later from Shell to build the Brent B. The initial contract for Beryl A was in the order of 330 million kroner (£23.4 million), while Brent B was estimated to cost 274 million kroner (£19.5 million). Beryl A was to be equpped with an integrated deck, while the Brent B deck would be almost empty.[17] The contractor promised to deliver the platforms in a time-span of 25 months. These export contracts were the largest ever registered in Norway.

The Condeep design is an offshore concrete drilling and production platform which rests on the sea-floor and maintains stability because of its weight, rather than being attached to the sea-floor by piles. It is also stable in the floating position at all times: during fabrication, towing and submerged. The lower section of the first Condeeps consisted of 19 circular cells in a honeycomb configuration, with a cell diameter of 20 m, and a height of 50 m, which came to rest on the sea-floor when the structure was submerged. These cells acted as the main buoyancy tanks during towing, but once placed at the offshore site they were the foundation of the structure and also provided oil storage. Access to the cells was provided by pumps and piping housed in one of the three columns. One leg served as the conductor through which the wells were to be drilled.

The lower part of the concrete structure was built in the dry dock. In the case of Beryl A (Figure 2) the work got under way in October 1973, Brent B following in November. The bottom sections were finished in the course of six months. When the lower part could be made buoyant, the dock was flooded and the partially completed structure was towed to a deep-water site in the fjord and moored. At this stage the structure contained some 50,000 m^3 of concrete and weighed in excess of 130,000 tonnes. Construction continued using the slipforming process to build the cell walls. This was a major undertaking requiring continuous operation. Pouring 1,700 m^3 of concrete a day for a period of 30 days was common. On a four-shift

Figure 2 Mobil North Sea Ltd. ordered the first Condeep production platform, to be placed on the Beryl Field. The lower part of the concrete substructure was built in a dry dock. The cell walls were erected through a continuous slipforming process in a floating position. The foundation of the structure consisted of nineteen cylindrical cells. In October 1974 sixteen of the cells were domed over, leaving three cells to be continued in the form of conical towers or shafts to support the deck and topsides.

Source: Norwegian Contractors.

basis this meant a work-force of 1,200 men. During slipforming the height of the cell walls above sea-level increased rapidly. Sea-water was pumped into the structure continuously, to ballast it down in order to keep the floating concrete mixing plant at the most efficient working level. After completion of the cell walls, sixteen of the cells were domed over. The installation of mechanical equipment for crude oil handling, drilling, ballasting, ventilation and instrumentation could commence at this stage.[18]

Erection of slipforms for the last step, the extension of three cells above water level to form conical towers or shafts to support the deck and the topsides, was done shortly after the completion of the domes. The deck section was to contain the drilling equipment, production and process equipment, as well as the living quarters and heliport for the platform. Constructing the conical shafts to a height in excess of 100 m was a major undertaking, requiring continuous operation for periods of approximately 40 days, and as much as 15,000 m^3 of concrete in one pour. By December 1974 the shafts had been completed on Beryl A, and by February 1975 on Brent B.[19] The final installation of mechanical equipment in the shafts continued throughout the spring months.

Gravity base structures are designed to remain upright during transit offshore. This makes it possible to install the topside facilities in sheltered waters before the platform is towed to sea. At the building site in Gandsfjorden, Stavanger, the water was more than deep enough, but the maximum tow-out draft between Gandsfjorden and the open sea was between 85 and 90 metres. Calculations showed that the first two Condeeps would be able to carry a load of up to 14,000 tonnes during tow-out. The accepted way of building topsides was to prefabricate supporting deck sections and prepackaged modules with maximum weights in the range of 1,500 and 2,000 tonnes, in line with the capacity of available offshore cranes. In the case of the new gravity platforms the complete supporting deck was to be floated over the totally submerged substructure, eliminating cranes and maximum weight limitations. When the deck had been mated with the substructure, the latter could be deballasted and conventional modules could be lifted on deck inshore. Shell chose this solution for Brent B.

An alternative would be to combine the supporting deck and the modules, including the processing plant, into an integrated deck, which would then be mated to the concrete structure. This would eliminate the overall need to use large and costly module-handling cranes, and drastically reduce the need for module hook-up inshore and offshore. The mounting costs of installing production equipment in modules after the platform was placed on the field made it attractive to install as much as possible before tow-out. The integrated deck concept seemed to give optimum utilization of the space on top of the platform.[20] Mobil chose the integrated deck solution for Beryl, which meant that structural steel designers and process/utility designers had to co-operate closely. In the case of the integrated deck the foundation for success was laid in the planning stage.

The Beryl A deck was assembled at a small Aker shipyard near Arendal. Two tanker hulls were placed on both sides of a pier to serve as supports for the deck assembly and later to carry the completed deck, weighing 12,000 tonnes, during 170 nautical miles of sea towage along the cost to Stavanger. The deck was assembled in the period from August 1974 to April 1975. Equipment and materials arrived late, and manpower peaked at approximately 1,500—more than twice the original schedule. Some of the work planned had to be delayed, and installations requiring yard facilities were given priority. When completed, the Beryl deck was three times as large as originally planned and construction drawings had been changed up to four times.

The deck was towed from Arendal to Stavanger in late April, and most of the remaining equipment was installed before the mating of deck and substructure took place. This was without doubt the most

Figure 3 The mating of deck and concrete substructure was the most critical part of the building process. On 30 May 1975 the deck on Beryl A was floated in the exact position above the three concrete towers. Only 6 m of the towers remained above the sea surface.

Source: Norwegian Contractors.

hazardous operation during the whole construction process. On 30 May 1975, the concrete structure had been ballasted down 120 m, with only 6 m of the shafts remaining above sea-level (Figure 3). The deck, resting on the tanker hulls, was then slowly towed into position over the transition pieces on the shafts. On 10 June the deck was in the exact position and the concrete structure deballasted until contact between the transition pieces and the deck took place. Deck and substructure were welded together before the deballasting continued, and the deck was lifted off the barges.[21] When the deck had reached a level safe from waves and other hazards, but low enough to reduce access problems for personnel, cranes and other types of lifting equipment, the deballasting was discontinued. In the following days and weeks the installation work proceeded at a brisk pace.

On Friday, 4 July 1975, the largest maritime towing operation ever undertaken got under way. Four tugs, combining 70,000 hp, went into action, towing the 350,000-tonne giant of concrete and steel out to the field, a distance of approximately 200 miles. On 8 July, less than two years after Mobil signed the contact with the Condeep Group, it arrived on location, at a water depth of 117 m. Thorough soil investigations of the conditions on the sea-bottom had been conducted long beforehand, and when the weather permitted, the substructure was ballasted and submerged onto the sandy sea-bottom. The foundation system incorporated steel skirts which were pressed down through the sea-bed to secure the platform against horizontal sliding.

Figure 4 View of concrete towers and deck from below, Brent B, June 1975.
Source: Norwegian Contractors.

The successful completion of the project was hailed as one of the major engineering achievements in the history of the offshore industry.[22] At the Offshore Technology Conference in early May 1975, before the mating of deck and substructure on the first Condeep had been completed, Høyer-Ellefsen was awarded a prestigious prize for pathbreaking innovations in the design of offshore production platforms. It was certainly a milestone in the history of Norwegian technology. The oil industry was impressed that both Norwegian Contractors and the Aker Group managed to fulfil the contract within the time limit, in spite of the fact that the project grew immensely in size from the planning stages to final delivery. Processing capacity was extended from 200,000 barrels a day to 300,000 barrels a day. Total costs for Beryl A grew from 330 million kroner to 1 billion kroner. The Condeep Group was also able to deliver Brent B in time. The deck was mated to the concrete structure in late June (Figure 4), and the platform was towed out to the Brent Field in August (Figure 5).

Figure 5 The second Condeep production platform, Brent B, under tow from the building site in Stavanger, Norway to the Brent Field east of Shetland in early August 1975. In the background to the left, work on the cylindrical cells of Statfjord A was going on.

Source: Norwegian Contractors.

A PASSING OR LASTING BREAKTHROUGH FOR CONCRETE GRAVITY PLATFORMS?

The first two Condeeps ordered in the summer of 1973 released a stream of orders for production platforms using concrete as the building material for the substructure. Oil companies signed contracts for twelve concrete gravity platforms between July 1973 and March 1975 to be installed on the British and Norwegian continental shelves. The majority were combined drilling and production platforms incorporating oil storage. All were for installation in the water depth range of 100–150 metres. Four different designs were used – Doris, Condeep, Sea Tank and Andoc, the latter three all using the same caisson and tower concept. Shell proved to be the company most willing to pioneer the use of concrete gravity platforms, by ordering five of the largest. The Norwegian Condeep design dominated the market, accounting for six of the fourteen platforms listed in Table 1.

Table 1 Concrete platforms installed in the North Sea in the 1970s.

Platform	*Design*	*Contractor*	*Ordered*	*Installed*
Ekofisk tank	Doris	Norw. Con.	May 1971	June 1973
Beryl A	Condeep	Norw. Con.	July 1973	July 1975
Brent B	Condeep	Norw. Con.	Aug. 1973	Aug. 1975
Frigg CDP1	Doris	Norw. Con.	Nov. 1973	Sept. 1975
Brent C	Sea Tank	McAlpine	Dec. 1973	June 1978
Frigg TP1	Sea Tank	McAlpine	Jan. 1974	Aug. 1976
Dunlin A	Andoc	Andoc	May 1974	June 1977
Cormorant A	Sea Tank	McAlpine	May 1974	May 1978
Brent D	Condeep	Norw. Con.	June 1974	July 1976
Frigg TCP2	Condeep	Norw. Con.	July 1974	June 1977
Statfjord A	Condeep	Norw. Con.	Aug. 1974	May 1977
Ninian	Doris	Howard/Dor.	Nov. 1974	May 1978
Frigg MP2	Doris	Dor./Skån.	March 1975	Aug. 1976
Statfjord B	Condeep	Norw. Con.	Feb. 1978	Aug. 1981

Concrete platforms came in on a wave of euphoria. They were seen for a short time as the only type of structure feasible in depths greater than 150 m in the northern North Sea. The British government gave active support to companies like McAlpine and Howard Doris in building the necessary facilities to construct concrete platforms at Ardyne Point and Kishorn in Scotland, while a string of other companies got permission to plan further building sites. Forecasts made in Britain in 1975 estimated that total demand for production platforms for installation in British waters by 1980 would be between fifty-two and sixty-four. On the supply side the builders might be able to produce up to eight steel platforms and eleven concrete platforms a year.[23] It was thought that demand would probably peak in 1977 or 1978.

The concrete platforms were viewed as very favourable solutions in the deepest waters. They had enough buoyancy to be floated out to location carrying a completed topside, saving the owners expensive offshore manhours and crane barge time. Engineers who were originally sceptical about concrete gravity platforms had by the end of the 1970s to concede that the design and the technology had proved their value under North Sea conditions.

However, the ideal of towing out a fully equipped drilling–production–storage platform from a fjord inshore to the field site offshore, with construction exposure to demanding and dangerous sea conditions minimized to a matter of days, was never fully achieved. Only to a certain degree did the various oil companies manage to optimize weight, size, cost and schedule. As a rule the oil companies had to live with both large cost and schedule escalations. The conclusion of both the British and the Norwegian independent study groups, appointed by their respective governments to report on the causes of escalation of cost of developing North Sea oil and gas, was that the offshore industry had underestimated the complexity of building permanent production installations in the new petroleum province.[24] It is fair to say, however, that the verdict on steel platforms was no more favourable than that on concrete platforms.

Nevertheless, the delays in field development worked in favour of steel rather than concrete. It was not the first time in the history of technology that an old technology was improved under the threat from a new one. During the late 1970s the industry witnessed an impressive evolution in steel-framed, pile-supported jackets for water depths up to 300 m. Builders of steel platforms showed an amazing ability to come up with the advances needed to keep pace with the growing demands. The designs and techniques for fabrication and installation in use in the mid-1980s greatly surpassed those used during the early days of North Sea operations. The power of piledriving hammers, the weight of jackets launched by barge and the size of offshore lifts increased by a factor of ten.[25]

During the 1970s pictures of huge Condeeps being towed out from the Norwegian coast to their respective locations by power tugs made the front pages and television screens not only in Norway and Britain but world-wide. Concrete gravity platforms became the symbols of oil exploration in the North Sea. But after the first rush to sign orders for concrete platforms, there was a prolonged period when no new orders materialized. While the Norwegian Condeeps were delivered on time, the ones built in Britain experienced costly delays, partly because of labour conflicts. Moreover, the shipping crisis in the mid-1970s came to have a strong impact on the production of production platforms. Shipowners cancelled contracts with shipyards, and order books that looked healthy in early 1975 were a dismal sight a year later. This led to large contractions in the European shipbuilding industry.[26]

Britain and Norway managed the crisis better than most other countries. In both countries the government intervened to reduce the size of the shipbuilding industry. But the managements at many yards, strongly encouraged by the authorities, shifted their attention from shipbuilding to the booming market for offshore structures. This made the necessary structural changes more gradual and less painful. On the other hand it led to growing protectionism in both the British and the Norwegian offshore industry. A pattern emerged where operators on the British shelf ordered steel jackets, while oil companies on the Norwegian shelf stayed with the Condeep.

The non-technological factors in the offshore context in the respective countries came to weigh heavily when oil companies came to choose between steel and concrete production platforms in the early 1980s. In Britain the companies building steel jackets had experienced fewer problems than the ones constructing concrete platforms. When demand levelled off and the contractors had to fight hard for the few contracts that came on the market, the oil companies felt more or less compelled to order steel platforms, and preferably from a British yard. The same was the case in Norway, but with reversed signs. Norwegian Contractors had successfully accomplished the building of the first Condeeps. Compared to the total cost of a complete production platform the concrete substructure was relatively inexpensive. The integrated steel deck and the mating of deck and substructure was the expensive part of the building project. The companies developing a Norwegian petroleum field could in theory order the deck from a foreign yard and tow it to the mating site in a Norwegian fjord. The mating operation and hook-up of modules would have to be done in Norway. This gave Norwegian firms a competitive advantage. Towards the end of the 1970s the different parties involved in the Norwegian oil industry knew that if an operator chose a Condeep solution, the contract for steel deck and mating would in all likelihood also be awarded to a Norwegian company.

In the 1980s the prevalent cultural context in Norway was geared to the survival of the Condeep, which in turn helped keep the Norwegian offshore industry occupied. The concrete structure by now was no longer an example of the technological style in the North Sea basin, but rather the Norwegian offshore style.

Notes and References

1. 'The North Sea Boils and Nearby Waters Simmer', *Drilling*, June 1965, 44–7; 'The North Sea: Stakes Are High in Oil's Biggest Gamble', *Offshore*, 21 June 1965.

2. 'How BP Views the North Sea', *Drilling*, July 1966, 88–9; 'British North Sea Activity is Soaring', *Offshore*, November 1966, 43; *North Sea Natural Gas* (London, Cazenove & Co., 1967), 7–9; F.G. Thackray, 'North Sea Gas:

How Much? How Soon? With What Consequences?', *Institute of Petroleum Review*, November 1966, 357–62.

3. The early activities on the Norwegian shelf and the Ekofisk discovery are discussed in detail by this author in Tore Jørgen Hanisch and Gunnar Nerheim, *Norsk oljehistorie: Fra vantro til overmot* [The History of Norwegian Oil], Vol. 1 (Oslo, 1992), 94–123. See also Stig Kvendseth, *Giant Discovery: A History of Ekofisk through the first 20 Years* (Stavanger, 1988), 19–32.

4. Daniel Yergin, *The Prize: The Epic Quest for Oil, Money and Power* (New York, 1991), 669.

5. 'North Sea Operators Back Optimism with Activity', *Oil and Gas Journal*, 8 January 1973, 70–2; 'North Sea Report: Overview', *Ocean Industry*, February 1973, 61–2.

6. Thomas P. Hughes, *Networks of Power: Electrification in Western Society, 1880–1930* (Baltimore, 1983), 405.

7. Griff C. Lee, 'Offshore Structures: Past, Present, Future', *Offshore*, 5 June 1968, 45–55.

8. 'Gas Platforms — How Designers Underestimated North Sea', *Offshore Engineer*, June 1975, 19–21.

9. 'Subsea Hardware, New Structure Types Coming to North Sea', *Oil and Gas Journal*, 8 January 1973, 83–92.

10. Hanisch and Nerheim, *op. cit.* (3), 190–1, 195–6.

11. Kvendseth, *op. cit.* (3), 51–3.

12. Hanisch and Nerheim, *op. cit.* (3), 200–3; Øyvind Steen, *I første rekke. Ingeniør F. Selmer A/S 1906–1981* (Oslo, 1981), 128.

13. Hanisch and Nerheim, *op. cit.* (3) 203–6; Kvendseth, op. cit. (3), 69–78.

14. For an overview of French contributions to offshore technology, see 'France — European leader', *Noroil*, May 1975, 97–102.

15. Hanisch and Nerheim, *op. cit.* (3), 252.

16. *Ibid.*, 253.

17. *Ibid.*, 253–5.

18. 'Mobil Beryl A — the Real Pioneer', *Noroil*, July 1975, 37–47.

19. Hanisch and Nerheim, *op. cit.* (3), 253–5.

20. Erik Holter and Knut Hagen, 'Integrated and Semi-integrated Deck Concepts for Production Platforms with Emphasis on Capital Cost Optimization', Offshore Technology Conference, OTC Paper 2686, Houston 1976.

21. Hanisch and Nerheim, *op. cit.* (3), 256–60.

22. 'Concrete and Protective Measures', *Noroil*, July 1975, 9.

23. G.A. Mackay and N.F. Trimble, *The Demand for Production Platforms and Platform Sites, 1974–80*, North Sea Study Occasional Papers, No. 3, February 1975, Department of Political Economy, University of Aberdeen. This study warned against the establishment of further building sites. In the course of a couple of years there would be excess capacity.

24. *North Sea costs escalation study*, Department of Energy, Energy Paper No. 7, (London, HMSO, 1976); *Kostnadsanalysen: Norsk kontinentalsokkel*, 2 vols, presented to the Norwegian Ministry of Oil and Energy, 29 April 1980.

25. 'How Steel Jackets have Kept Taking the Hurdles', *Offshore Engineer*, April 1984, 80–8.

26. Bo Stråth, *The Politics of De-industrialisation: the Contraction of the West European Shipbuilding Industry* (London, 1987).

Edward A. Uehling and the Automatic Pig-Casting Machine: A Case-Study of Technological Transfer

W. DAVID LEWIS*

The utility of a technological innovation depends heavily upon how it fits into the cultural environment in which it is introduced. A case-study supporting this generalization is provided by the development and reception of a machine that was invented in the late nineteenth century by an American engineer, Edward A. Uehling, to cast pig iron automatically in individual moulds. Conceived in Birmingham, Alabama in the early 1890s, Uehling's invention was not adopted there because the economic and social conditions under which pig iron was made in the southern part of the United States at that time were not well suited to the mechanization of blast furnace practice. Conditions in western Pennsylvania and other northern centres of iron and steel manufacture, however, were favourable for utilizing the invention. Consequently, it became a key feature of the 'Duquesne Revolution' that began in the Pittsburgh District during the 1890s.[1] The automatic pig-casting machine was also adopted by a number of European firms, but the contrast between its reception in Great Britain on the one hand, and the European continent on the other, was consistent with what Uehling had experienced in the United States. This contrast revealed once again the impact of economic and social circumstances upon technological tranfer.

Born to German immigrant parents in Dodge County, Wisconsin, in 1849, Uehling displayed mechanical precocity at an early age. Given the best education that his father, a farmer, could provide from limited earnings, Uehling studied for a year at Northwestern College in Watertown, Wisconsin, excelling in mathematics and physics. In 1873, he entered Stevens Institute of Technology at Hoboken, New Jersey, after reading about its engineering curriculum in *Scientific American*. His work at Stevens impressed the president of that institution, Henry L. Morton, whose support was critical to the subsequent development of Uehling's career.

After receiving his mechanical engineering degree at Stevens in 1877, Uehling stayed briefly at the school to assist Henry Thurston, one of America's most distinguished engineering educators. Under a commission from a Pittsburgh firm, Jones & Laughlin, the two men collaborated in doing research on cold rolled iron and structural steel. In 1879, Uehling left Stevens to become master mechanic for the Douglas Furnace Company of Sharpsville, Pennsylvania, where he gained practical experience in blast furnace operations. In the 1880s, he managed five blast furnaces for the Bethlehem Iron Company. Early in 1890, he was disappointed when that firm's acceptance of a government order for armour plate led it to build an open-hearth steel furnace and cancel plans under which Uehling was to build a battery of hot blast stoves. Soon afterwards, Uehling decided to accept a new post at the Sloss Iron and Steel Company in Birmingham, Alabama.[2]

Sloss had been in turmoil ever since its founding in 1881. Under its original owner, Colonel James W. Sloss, it had failed to acquire its own coal and ore mines, resulting in costs that were higher than those of most other firms in the Birmingham District. In addition, the two original furnaces that Colonel Sloss built in 1881–3 were poorly designed for utilizing local raw materials. New owners who took control in 1887 obtained independent access to coal, ore, and limestone, but continued to struggle with other problems, particularly a shortage of skilled technicians in an urban frontier area whose rapid growth had outstripped its human and financial resources. As of 1890, the company operated four blast furnaces, including two new stacks that had been built at North Birmingham, an industrial community recently created by the new owners. None of the four furnaces, however, was working properly. Most of the firm's directors were railroad promoters from Richmond, Virginia, and knew little about technical operations. There was much friction among them about how to deal with the unprofitable state of the enterprise. The problems with which they grappled were partly traceable to the managerial inexperience of the company's president, Thomas Seddon, who had no background in iron manufacturing. Seddon also had a strained relationship with Sloss's chief stockholder and dominant director, Joseph Bryan, a wealthy entrepreneur who lived in Richmond.[3]

Bryan and his associates hired Uehling because of the favourable impression he had made on Henry Morton during his undergraduate days at Stevens. Morton was one of Sloss's few northern directors, and Bryan had sought his advice as the company's performance deteriorated. Morton believed that Sloss had a potentially bright future under proper management. On his recommendation, Seddon was relegated to a largely titular role as president, and two of Morton's former protégés at Stevens were hired to take charge of operations. William Dashiell, a Marylander who became general manager at Sloss, was one of these. Uehling, who became blast furnace superintendent, was the other.[4]

In Alabama, Uehling quickly found himself in a hornets' nest.

Seddon, who was determined to regain his former power, was popular among Birmingham investors, who resented the authority wielded by Bryan and other absentee owners. Within a few months, Seddon demanded Dashiell's dismissal, causing a crisis among the directors. At a board meeting held in Birmingham late in 1890, Dashiell was forced to resign. Seddon, however, needed to get the company's furnaces working properly and lacked the background to accomplish this result. Because of his lack of expertise, he promised Morton that he would co-operate with Uehling if the latter stayed on as furnace superintendent. Uehling therefore remained in Birmingham and gradually restored the furnaces to good operating condition.[5]

Figure 1 Sand casting of pig iron by hand at a blast furnace typical of the ones operated by the Sloss Iron and Steel Company at the time Uehling worked in Birmingham.

Source: Edward A. Uehling: 'Pig-Iron Casting and Conveying Machinery', *Cassiers Magazine*, 1903, 24:115–18.

As was true throughout the industry at the time, pig iron was moulded in beds of sand in a casting shed located directly in front of a blast furnace (Figure 1). None of the work was mechanized. Gushing from the tapping hole with explosive force, showering the casting shed with sparks that burned the flesh of workers who did not quickly get out of the way, molten iron ran into a central trench, the 'runner', after which it flowed into lateral channels, known as 'sows'. These channels led in turn to smaller moulds, known as 'pigs', in which individual billets of iron were formed. Cast-iron stops, called 'shutters', were manually opened and closed to make sure that each cluster of pigs filled evenly. Gradually solidifying, but still red-hot, the iron

Figure 2 Separating pigs from sows with sledge-hammers in a casting shed, typical of the onerous work that offended Uehling's sensibilities.

Source: Edward A. Uehling, *op. cit.* (Figure 1).

was covered with layers of sand by workers who wore protective shoes with heavy wooden soles as they separated pigs from the sows with sledge hammers (Figure 2). After being sprayed with cooling water, the pigs, weighing up to 125 pounds each, were hand-loaded onto mule-drawn carts and taken to the furnace yard for shipment (Figure 3).[6]

As Uehling later noted when he applied for a patent on his automatic pig-casting machine, the traditional method that has just been described resulted in extremely uneven output. Up to fifteen grades of iron, ranging from 'No. 1 Extra Selected' at the top to 'White' on the bottom, were identified after testing by fracture. 'The product is rough and irregular and the operation is very wasteful, producing large quantities of scrap, which must be remelted, and

Figure 3 Hand-loading cooled pigs on mule-drawn carts to be taken to the furnace yard for shipment.

Source: Edward A. Uehling, *op. cit.* (Figure 1).

carries much sand with the iron, which reduces the value of the iron', Uehling wrote. 'Moreover, this method of running the iron from the furnace into the sand is under very poor control. It sometimes runs so slowly that the iron chills in the long runner before it reaches the lower pig-beds. At other times the running of the iron is with such a rush that the cores, which form the partitions between the pig-moulds, are washed away . . . In either case the iron must be taken up from the sand beds and remelted to make it marketable, besides the expense of having to break it up and remove it from the beds.' Because some parts of a melt had too little silica and graphitic carbon and others too much, two or more grades of iron might result from the same melt.[7]

Bad working conditions also affected the quality of the product. At every stage, the work was hot, arduous and dangerous. The only thing that made Sloss and other southern furnace companies distinctive among other American ironmaking enterprises was their use of black workers in all but supervisory positions. African-Americans were recruited from impoverished rural areas, received low wages, and were subjected to severe social disincentives that compounded the problem of securing proper levels of performance and discipline. Inefficiency and chronic absenteeism forced employers in the Birmingham District to hire at least twice as many workers as they ideally should have needed to staff their installations.[8]

None of the furnaces under Uehling's supervision was more than 75 feet high. Their individual output did not normally exceed about 200 tons of iron per day, and melts ranged from 40 to 50 tons each. Even so, with his inefficient and unmotivated workers, Uehling found it hard to get a casting floor cleared of pigs in time for the next melt to be released from a furnace. As a result, he conceived a new approach, aimed at mechanizing the casting process. Uehling was not motivated solely by economic and technical considerations. His writings manifest a deep moral earnestness, and show that he was emotionally repelled by the sheer onerousness of work in the casting sheds. 'The task of breaking and carrying out the iron from the casting beds of even a moderate sized furnace is not a fit one for human beings', he stated. 'If it were possible to employ horses, mules, or oxen to perform this work, the Society for the Prevention of Cruelty to Dumb Beasts would have interfered long ago, and rightfully so.'[9]

Seeking both to improve efficiency and alleviate drudgery, Uehling devised an automatic system of pig-iron casting that he worked out on paper but never implemented in practice during his years in Birmingham. His invention was elegantly conceived and is all the more remarkable because he had no means of making a prototype or model of his apparatus. The large size of the equipment required for his casting system was essential to its proper functioning; building a prototype would have necessitated space, materials, and levels of funding that the Sloss Company could not possibly have given Uehling, even had its owners and managers been sympathetic to his

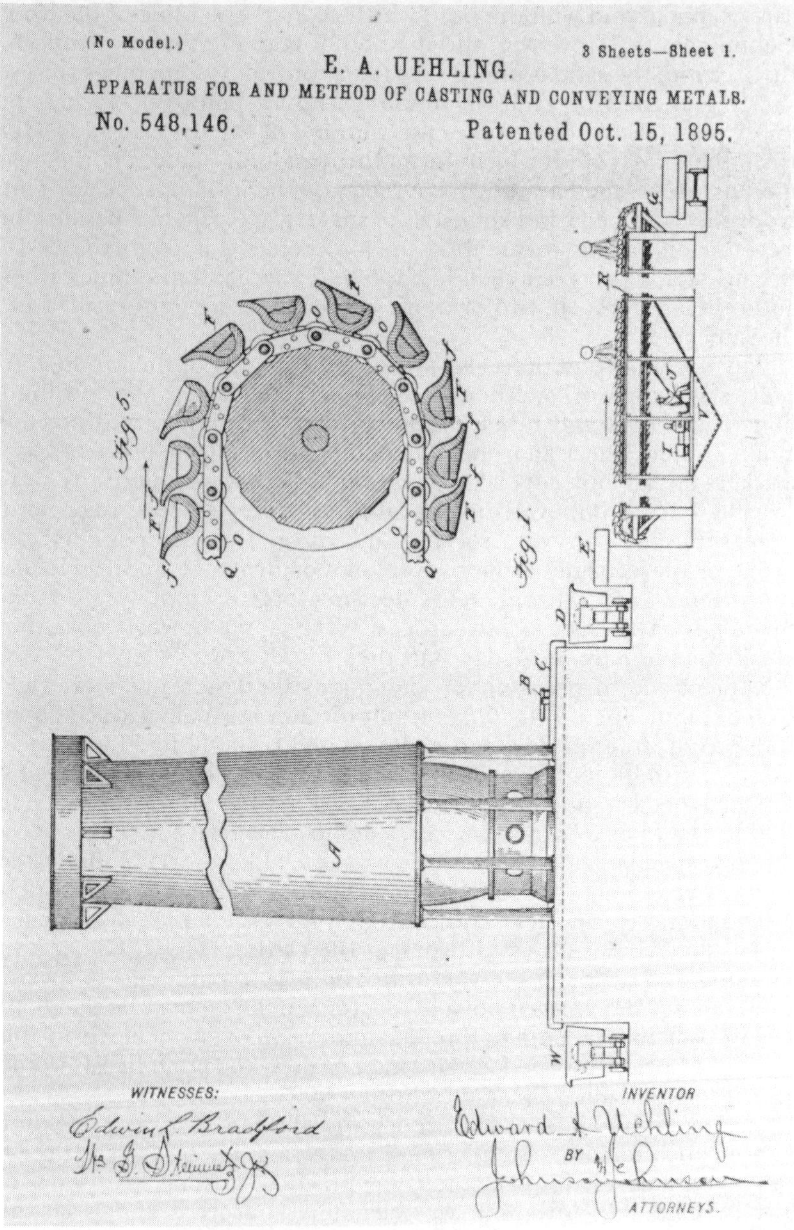

Figure 4 Scale drawing relating to Uehling's automatic pig-casting machine, from patent specification of 15 October 1895.

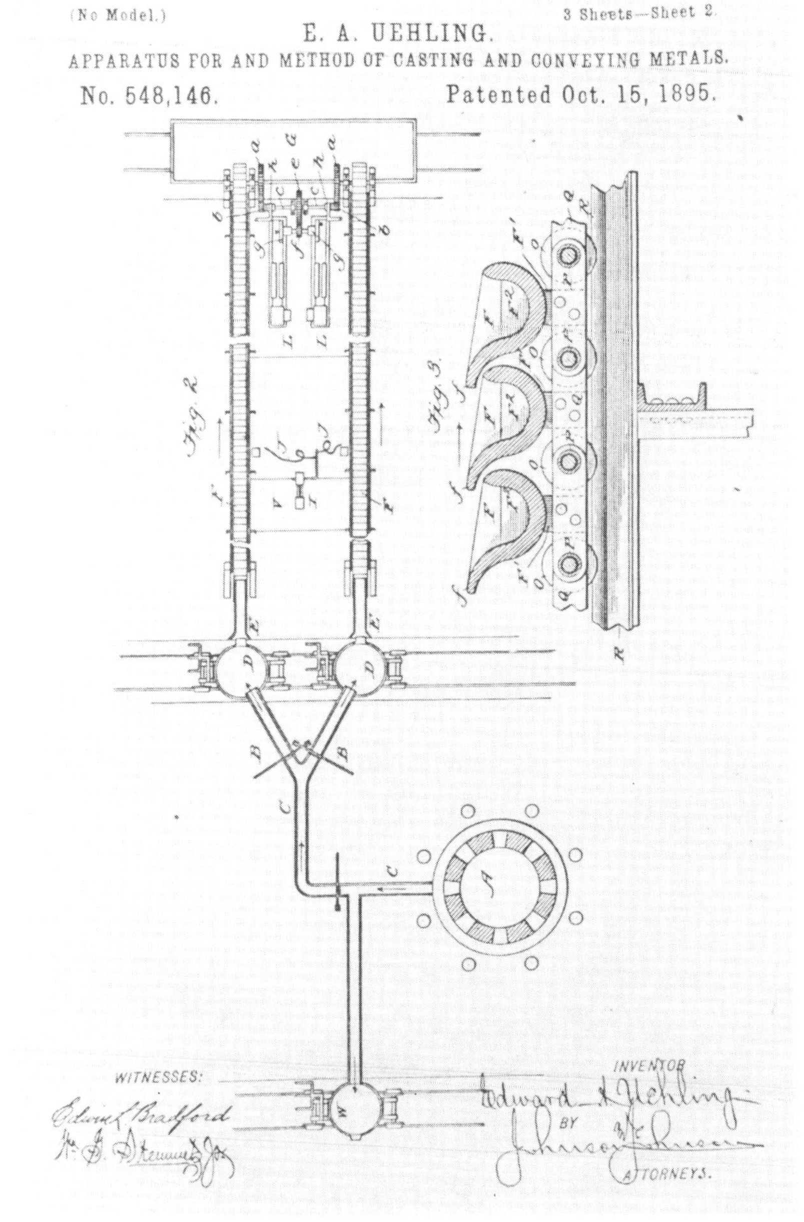

Figure 5 Scale drawing relating to Uehling's automatic pig-casting machine, from patent specification of 15 October 1895.

ideas and purposes. That the machine existed in such detail only in Uehling's mind until after he left Birmingham makes his inventive genius self-evident.

In Uehling's system (see Figure 4), molten iron from a blast furnace (A) was tapped through a runner (C) into a large mixer of 40 to 50 tons capacity (D). Because the surface area of the iron in the mixer was small by comparison with what it would have been had the iron been distributed throughout a maze of channels in a casting-bed, the iron remained hot for several hours and could 'be treated in any manner that may seem expedient to improve its quality or change its character before tapping it into molds,' as Uehling stated. After the iron had been mechanically stirred and blended, the mixer was tilted forward and its molten contents were slowly discharged through a spout (E) into an endless chain of bucket-shaped iron moulds (F). In Uehling's original drawings, each mould was partly divided by a transverse rib that enabled two pigs to be cast at once in the same travelling container, connected by a thin coupling so that they could be easily knocked apart Figure 5: F, F^2). This, however, was not a necessary feature of the apparatus.

The moulds moved forward on a carrier made of interconnected links that were held together by axles (Figure 5: Q) whose projecting ends terminated in flanged wheels (O) that ran on rails situated on either side of the links (R). The rails were mounted in a supporting structure so that the flanged wheels would be held in place regardless of whether the moulds were facing upward or downward. At each end of the structure were drum-shafts that were connected to an impelling engine. The system was designed so that two or more strands of mould-conveyors could be operated simultaneously. Every detail of the drive mechanism and of the moulds themselves exemplified the care that Uehling took to make sure that the equipment functioned

Figure 6 Delivery end (left) of single-strand Uehling casting machine installed by Struthers Furnace Company, Struthers, Ohio.

Source: Edward A. Uehling, *op. cit.* (Figure 1).

smoothly. For example, the rear side of each individual mould was curved so that the lip (f) overlapped the front of the mould immediately behind it (Figure 5). This prevented spillage of molten iron that would have otherwise have dripped down onto the endless drive mechanism that was moving the moulds forward, thus rendering the machinery inoperable.

Because the travelling moulds were open to the ambient air on all sides and were completely exposed to it at the top, heat in the molten iron that they were carrying dissipated much more rapidly than was possible in beds of sand on a casting floor. Depending on the rate of speed at which the machinery was operated, the contents of the moulds might solidify within a distance as short as 50 feet. The interior surfaces of the moulds were coated with graphite in order to prevent them from becoming welded to the molten iron that they were carrying.

Figure 7 Pouring end (centre right) of Struthers Company's machine.

Source: Edward A. Uehling, *op. cit.* (Figure 1).

Travelling upwards and cooling as it moved, the metal in the moulds solidified into billets of pig iron. As the iron began to solidify, a water spray hastened the cooling process and helped promote shrinkage of the iron so that it would separate more readily from the mould. As the endless conveyor reached a revolving drum at a point opposite from the mixer, the moulds began to tilt downward and the iron fell into a waiting rail car. As they were emptied, the moulds became inverted for the return journey to the mixer. Now facing downward, the moulds passed above a reservoir filled with a mixture

Figure 8 A six-strand Uehling casting machine at Andrew Carnegie's Edgar Thomson steelworks, epitomizing the scale of production in western Pennsylvania.

Source: Edward A. Uehling, *op. cit.* (Figure 1).

of graphite and water (Figure 4: V). (Later applications of the system used a solution of water, crushed limestone, and powdered coal.) In Uehling's original design, the water constituted a medium to convey graphite from a spray rig against the interior surfaces of the inverted moulds. Using a pressurized hose (see pump shown in Figure 4:I) with a nozzle that was specially designed to distribute the graphite–water suspension evenly, a worker (J) sprayed the moulds as they passed overhead. Excess graphite and water splashing away from the moulds hit a dashboard that channelled these ingredients back into the reservoir to be reused, thus preventing waste. Heat radiating downwards from iron in moulds that had still not reached the point where dumping occurred quickly caused water in the freshly sprayed moulds to evaporate, so that only the graphite remained. As the inverted moulds approached the mixer on the return journey, they passed under progressively hotter moulds that had just been filled with molten iron. The downward radiation of ever greater amounts of heat from above ensured that any water vapour remaining in the inverted moulds would be eliminated by the time these moulds reached a revolving drum that would bring them into an upright position. At that time, they would be refilled with molten iron from the mixer and begin another circuit.[10]

Uehling's system was a beautifully conceived assembly of components skilfully integrated to achieve the automatic mechanical casting of pig iron in identical units with uniform chemical characteristics. Uehling patented the process in October 1895 but never had an opportunity to implement the system in Alabama. Because of his northern background and the circumstances surrounding his appointment, Seddon had never liked or fully accepted him. In 1895, after Uehling had repaired the company's furnaces and secured satisfactory levels of performance, Seddon demanded his resignation. Complying with Seddon's wishes, Uehling remained temporarily in Birmingham to develop a pneumatic pyrometer that accurately measured the heat of the air supplied to a blast furnace by its stoves. After patenting the device, he left Alabama in 1896 to establish a shop in Newark, New Jersey, where the pyrometer could be commercially produced. Meanwhile, Morton, who was increasingly disenchanted with the way in which Sloss was being managed, withdrew from the firm.[11]

Not until the late 1920s did Sloss adopt Uehling's pig-casting methods. The reasons why the company rejected his system in 1895, however, had nothing to do with Uehling's poor relationship with Seddon. Sloss's chief investor and dominant force, Joseph Bryan, admired Uehling greatly, was reluctant to see him leave Alabama, and could have spearheaded utilization of his machine had he so desired.[12] The reason Bryan did not do so was because Uehling's device did not fit the economic and social presuppositions underlying blast furnace practice in the Birmingham District.

Like all the entrepreneurs who owned and operated blast furnaces in and around Birmingham, Bryan and his associates were committed to an underlying strategy for competing with pig iron producers outside the region. This strategy was based in part on the fact that Birmingham was the only place in the world that had all three of the raw materials that were needed to make pig iron — ore, coal and limestone — available in great quantities within a thirty-mile radius of Jones Valley, in which the city was located. Capitalizing on this unique situation, owners acquired 'captive' mines and quarries that were in some cases only a few miles away from their furnaces. This resulted in extremely short lines of supply, cutting costs. Some companies also owned their own rail lines, yielding further cost benefits. By marshalling raw materials as efficiently as possible, furnace owners practised a form of vertical integration that was locally known as 'straight line production'. As one analyst aptly stated in 1924, 'the making of iron in the Birmingham District is first, last and always a great material handling business'.[13]

Mechanization was less an integral part of 'straight line production' than a supplement to it, to be used only when needed. This was due to another aspect of Birmingham's competitive strategy, the use of low-paid black labour. Despite the inefficiency of its badly disincentived workers, Sloss did not mechanize casting and other furnace operations until the late 1920s and early 1930s. Even the Woodward Iron Company, which was by far the most profitable firm in the entire District, delayed adopting mechanical casting until about the same time. In part, this was because the foundry trade, to which both Sloss and Woodward catered, had a long-standing preference for sand-cast pig iron as opposed to the machine-cast variety. But the persistence of labour-intensive methods was also traceable to the compatibility of this policy with low costs in the South, where disadvantaged black workers were easy to recruit. During the late 1880s, just before Uehling moved to Birmingham, pig iron could be made in the District for less than $12 per ton.[14]

Bryan and other southern iron manufacturers were committed to a labour-intensive policy partly because it did not demand the substantial amounts of capital that would have been needed to implement an approach involving a high degree of mechanization. Southern blast furnace practice contrasted with a machine-intensive strategy that northern entrepreneurs had begun to pursue as early as the 1870s, when Pittsburgh's Lucy and Isabella furnaces first began to make pig iron in unprecedented amounts, using a technique known as 'hard driving'. Besides involving blast pressures unsuited to the lean ore and relatively soft coke of the Birmingham District, this technique was also too costly to utilize in the South, because that region had to depend largely on its own capital to fund manufacturing enterprises throughout most of the late nineteenth century.[15] Even though the region had many friends on Wall Street, which became the major source of funding for large business ventures during that period, banks and brokerage houses headquartered in New York City did not tend to risk money in industrial securities. Until the turn of the twentieth century, most of their investments in southern development went into railroads.[16] Typifying this state of affairs, Sloss's board of directors included few northerners; Bryan, a Confederate veteran who was strongly committed to the idea that the defeated South could compete on approximately equal terms with northern industry, was careful to keep control of the firm in southern hands.[17] Raising the funds needed to adopt Uehling's invention, even in the unlikely event that the money could have been found, would have run counter to this policy.

Another reason why Uehling's device was ill-suited to conditions in Birmingham was that markets for the District's pig iron were inadequate to support levels of production that would have made mechanical casting practical. This situation was due partly to a dearth of foundries and rolling mills in the South. Most of the pig iron that was produced in the region had to be exported to other parts of the country to be made into finished or semi-finished goods. An even greater problem than restricted markets, however, was the chemical composition of the raw materials available in the Birmingham District, which did not permit the manufacture of pig iron that could be converted into steel. Besides being extremely lean, ore deposits in Birmingham's famed Red Mountain contained large percentages of phosphorus and silica. Meanwhile, coal from the nearby Warrior Field was high in sulphur. All of these chemical elements were incompatible with the acid Bessemer steelmaking process that spread from Great Britain to America in the 1860s and remained dominant in the United

States throughout the late nineteenth century. European and American inventors eventually solved this problem by developing the basic (as opposed to acid) open-hearth process, but this was only starting to be used in the United States when Uehling conceived his pig-casting device.[18]

As a result of the raw materials on which they were forced to depend, none of Birmingham's furnace companies made steel in commercially viable quantities for almost three decades after the city was founded in 1871. The corporate name of the Sloss Iron and Steel Company was therefore highly misleading. It had resulted from an abortive attempt made by Colonel Sloss in the mid-1880s to obtain disputed patent rights from a northern inventor, Jacob Reese, to produce basic steel. Neither Sloss nor its successor firm, the Sloss-Sheffield Steel and Iron Company, ever made an ounce of steel.

Just before the twentieth century opened, the first significant quantities of basic steel ever made in Birmingham were produced after an arduous and costly programme of research and development by the District's largest firm, the Tennessee Coal, Iron, and Railroad Company. Known locally as TCI, this enterprise had moved to Alabama from Tennessee in 1886. Until 1899, however, every effort to manufacture steel in Birmingham had resulted in failure, and the District produced only pig iron. This commodity, far less profitable than steel, was shipped to such nearby destinations as Cincinnati, Louisville, and St Louis, where foundries and rolling mills converted it into wrought iron to make agricultural implements, boiler plate, and other specialized products. The only other place besides the lower Ohio Valley where southern pig iron enjoyed a competitive advantage was in urban centres along the north-eastern seaboard, where it vied successfully with the output of firms in eastern Pennsylvania that utilized obsolete anthracite-fired furnaces. Because their markets were so limited, Birmingham-based companies could not aspire to production levels as high as those maintained by blast furnaces in the Pittsburgh District, which supplied pig iron to sprawling mills that made a highly standardized and extremely profitable product, steel rails, in huge quanities. Because of the limited markets for the Sloss Company's pig iron, it would have made no sense for Bryan and his associates to invest scarce capital in mechanization to maximize output.[19]

It is worth noting, however, that Birmingham's furnace operators did practise a *selective* aproach to mechanization, and other forms of modernization as well, when it suited their purposes to do so. Because of the sulphur and shale that abounded in Warrior coal, for example, the Birmingham District led the nation in adopting mechanical coal-washing, which was necessary for local beehive ovens to make usable grades of coke. Later, partly because by-product coke was better suited than beehive coke to the relatively small size of Alabama's blast furnaces, and also because hard ore mined from deposits deep under Red Mountain demanded a stronger type of coke then beehive ovens could produce, Alabama became the first American state to abandon beehive ovens in favour of by-product models. Similarly, when geological conditions made it imperative to do so, local furnace companies adopted innovations in coal and ore mining that were comparable to the most up-to-date techniques utilized elsewhere. In the early 1890s, at the same time that Bryan and his fellow directors rejected Uehling's pig-casting device, they installed modern coal-cutting machinery at a mine in Brookside, Alabama, because the nature of the strata in which the coal was imbedded made hand-picking extremely difficult. Only under such conditions did a careful, piecemeal approach to mechanization make rational sense.[20]

A discriminating approach to mechanization was also appropriate in Birmingham because, like the South in general, the District depended heavily on technical assistance from outside the region. In the *ante-bellum* era, slaveholding industrialists had made a practice of recruiting northern and foreign engineers, laying the foundations for widespread use of the same approach after the Civil War. Such help, however, was obtainable only at high cost. Uehling's move to Birmingham was a typical case in point. He came to Sloss for $5,000 per year, a salary one-third higher than that which he had earned at the Bethlehem Iron Company. A similar policy of hiring high-priced technical personnel was followed by TCI, which, three years before Uehling arrived in Alabama, had lured Erskine Ramsay, a highly creative northern engineer of Scottish descent, to come to Birmingham. Ramsay came to the District from western Pennsylvania's famed Connellsville coke-producing region because TCI offered him substantially higher pay than he could obtain from his previous employer, coal baron Henry Clay Frick. Staying in Birmingham for the rest of his long and distinguished career, Ramsay patented a series of mechanical devices, including coal-washers, revolving dumpers, mine-car couplers, slag-pot cleaners, and various types of sampling and testing apparatus, that significantly enhanced the way in which TCI utilized coal and other raw materials.[21] Importing such engineers as Ramsay enabled the Birmingham District to stay abreast of the best outside practice in certain specialized types of operations. But the lack of southern-born engineers, the scarcity of southern capital, and the high cost of securing outside help did not permit the District to emulate the degree of mechanization that became common in the North.

Uehling's failure to win adoption of his machine in Alabama was also due in part to deep-seated racial attitudes prevailing among southern owners and managers, who did not share his moral and emotional reservations about the gruelling work demanded of black operatives in the casting sheds of the Birmingham District. Despite being a prominent symbol of the New South, Birmingham was thoroughly steeped in Old South traditions. The land speculators who had planned the city were mostly connected with the 'Broad River Group', an association of wealthy planters who had owned large numbers of slaves before the Civil War.[22] Joseph Bryan, who had been raised on a Virginia tobacco plantation, Eagle Point, had a similar background. As one account of his career states, he epitomized a tradition of 'economic orthodoxy, elite rule, lower-class deference, and black subservience'.[23]

Partly because of a social stigma that many southern whites attached to occupations that they regarded as menial, industrial slavery was common in the South before the Civil War. Ironmaking itself was a plantation-type industry in *ante-bellum* times, and all of Alabama's pre-war blast furnaces were built and operated by slaves. Such use of servile labour was compatible with selective use of machines. Louisiana was one of the first American states to use stationary steam engines on a large scale; sugar planters installed them in slave-run mills where cane was processed after harvesting. Pre-war Richmond was a hive of slave-operated industries; it was famed for such enterprises as the Tredegar Iron Works, where owner Joseph R. Anderson used chattel labour to operate rolling mills. Wherever possible, however, industrial slaveholders used labour-intensive methods in preference to a northern emphasis on mechanization that was already emerging before 1861. Reliance on manual labour, for example, was highly characteristic of southern agriculture. Wealthy cotton planters reaped handsome profits by using large numbers of enslaved field hands while northern wheat farmers adopted such mechanical devices as McCormick's harvesters, originally invented by a southerner from Virginia. It is not surprising, therefore, that post-war furnace owners in the Birmingham District, who came from agricultural backgrounds and shared the social views of their *ante-bellum* predecessors, continued to rely on large numbers of low-paid black workers as an alternative to a northern model of industrialization based on intensive use of labour-saving devices.[24]

Late in the *ante-bellum* period, when it was first envisioned, Birmingham had been conceived as an 'urban iron plantation'. As indicated in a prospectus written in 1859 by one of Birmingham's founding fathers, John T. Milner, the city's blast furnaces would have been manned by slaves but for the fact that the South lost the Civil War. By the time Birmingham was founded in 1871, slavery was illegal, but the District's industrial leaders continued to manifest a strong preference for servile black labour. A large part of their work-force was composed of state and local convicts, mostly African-Americans, who mined coal and ore under a leasing system that was not abandoned until 1928. Much of the rest of the District's labour supply consisted of blacks who were recruited by subcontractors of the same race, continuing another practice inherited from *ante-bellum* times. The system was predicated on the belief that blacks were best suited for hot, dangerous and dirty occupations requiring brute strength, and that using them in this way left whites free to engage in other, more desirable types of work. Milner, who had surveyed one of Birmingham's main rail connections to the outside world, had shown a strong preference for using black workers in railroad-building before the war. After the conflict ended, he and other southern industrialists were confident that they had a unique ability to discipline and get the most from such a labour force. As Milner frankly admitted, this belief was a key element in the faith held by southern manufacturers that they could compete successfully with northern businessmen.[25]

As a result of the plantation-bred mindset prevailing among its founders and furnace owners, Birmingham became atypical, even in the South, for the percentage of its population consisting of African-Americans. The number of its black residents, starting with 2,506 in 1871, doubled by 1880, increased sixfold by 1890, and continued to grow steadily until it reached 130,000 in 1920. At the time Uehling came to the city in 1890, 43 per cent of its inhabitants were black. In time, Birmingham became known as the 'Johannesburg of America'. Photographs taken during the late nineteenth century showing conditions in the two cities, one in north-central Alabama and the other on the South African veldt, are eerily similar, showing large groups of black workers taking part in a variety of industrial occupations. As of 1886, 90 per cent of all workers at Birmingham's blast furnaces were black; the remaining 10 per cent were white supervisors. This policy avoided mixing the races as equals in the workplace, a practice that southern entrepreneurs abhorred. Depending on this type of workforce also made effective use of surplus black sharecroppers at a time when sagging cotton prices had created a glut of agricultural labour. However bad conditions might be in Alabama's mining camps and urban slums, they were much worse in the state's wretched rural districts. Given the shortage of southern capital and the lack of markets for high-quantity production, the vast pool of black workers on which Birmingham's industrial leaders could draw, when added to the prevalence of white supremacist values, made Birmingham's business leaders all the more heavily committed to a labour-intensive, not a labour-saving, strategy. There was no chance for Uehling's pig-casting machine to be adopted in such a social and economic environment.[26]

Conditions were much different in the North, especially in the Pittsburgh District. Because historians have repeatedly analysed this sector of the American ferrous metals industry in great detail, while paying much less attention to developments in the South, the general features of northern practice can be quickly summarized. From the 1870s onwards, blast furnace operators in western Pennsylvania became increasingly committed to spending vast amounts of money in transporting huge quantities of raw materials from places as far away as the Lake Superior region to enormous installations in the Pittsburgh District's Monongahela Valley. There, a lavish expenditure of capital on machine production reduced unit costs of production to the lowest possible extent. Unlike southern raw materials, the iron ore and coal available to northern furnace operators from the upper

Michigan peninsula and the Connellsville District of western Pennsylvania resulted in pig iron that was extremely suitable for the acid Bessemer technique of steel manufacture. This high degree of adaptability enabled Andrew Carnegie and other industrialists to integrate blast furnaces with heavily mechanized mills that could capitalize on a high volume of demand for steel rails at a time when the nation's railroad system was constantly expanding.[27]

Such circumstances provided an appropriate setting for Uehling's pig-casting machine. Enormous blast furnaces, capable of making up to 600 tons of pig iron per day for such sprawling installations as Carnegie's Edgar Thomson, Duquesne and Homstead steel works, fitted logically into the northern style of practice based on large-scale production of ferrous metals. As Uehling stated, a furnace burden of 250 tons could be cleared from a casting shed only 'with great difficulty' before the next melt took place. When burdens exceeding 300 tons were reached, 'the situation became desperate. Something had to be done.'[28]

An abundance of capital and the existence of a vast market for standardized products were not the only things that made the Pittsburgh District an ideal place for intensive, as opposed to selective, mechanization. Because Pennsylvania and Ohio had long been the chief bastions of ferrous metals production in America, Carnegie and other entrepreneurs could draw on an abundance of highly experienced engineers and technicians who understood every aspect of the iron and steel industry. Operators in the Pittsburgh District also had convenient access to a wide variety of machine-manufacturing enterprises that had long made advanced industrial equipment. Enjoying ideal raw materials and market opportunities that demanded an unrelenting effort to maximize output, Pittsburgh's industrial leaders needed to eliminate as many production bottlenecks as possible. Given the capital at their disposal and all of the other advantages just mentioned, they had every opportunity to do so with ease.

Labour costs were of more consequence to northern operators than was true in Alabama, because the traditional northern commitment to a free labour system resulted in much higher wage levels than those prevailing below the Mason and Dixon Line. Due to the long distances separating Pittsburgh from some of its raw materials, particularly iron ore, and also because steel production required much greater skills than mere pig iron manufacture, labour was a more critical variable in Pittsburgh than it was in Birmingham, where all necessary raw materials were close by and little skilled labour was needed. As labour historian David Brody had stated in his classic study, *Steelworkers in America*, 'every effort was directed to lowering the labor cost per ton of steel' in western Pennsylvania. Inevitably, the need among Pittsburgh's industrialists to pay close attention to labour costs as a key variable affected blast furnace practice as well as operations in the steel mills themselves. Having ready access to capital, technical expertise, and local machine works that could meet any conceivable need for advanced equipment, Carnegie and other owners naturally opted for intensive mechanization as the best way to facilitate the high-volume operations that their market opportunities demanded. As Brody indicates, 'reduced labor costs . . . were achieved by attention to machines, not labor'. Not only in steelmaking operations but in pig iron production as well, as Brody has stated, 'hordes of common laborers were rapidly supplanted'.[29] By the 1890s, therefore, northern and southern systems of blast furnace practice had become two radically different worlds, based on contrasting assumptions.

Even before Uehling left Alabama, Carnegie learned about what he was doing and 'sent a representative to Birmingham to negotiate for the purchase of my patent,' as Uehling later wrote. Uehling asked $50,000 for the right to utilize his machine at any place in the United States; $20,000 for its use in Pennsylvania's Allegheny County, in which Pittsburgh is located; or $10,000 at one single furnace. Ultimately, he settled for $2,000 from Carnegie for a ninety-day option to test the device at the latter's Lucy Furnace. Lucy's superintendent, James Scott, was already thinking along lines analogous to what Uehling had been doing at Birmingham. Scott, however, was pursuing a method that Uehling described as 'even more radical than mine'. Scott's technique involved pouring a thin stream of molten pig iron into water, causing it to granulate, and converting the resulting granules directlty into steel. As Uehling saw, the technical problems involved in adapting such a process to large-scale mechanized production were formidable. The abrasiveness of granular iron imposed great wear and tear on the equipment utilized by Scott. In addition, owing to an extremely high ratio between surface area and weight, granular iron oxidised rapidly and hence could not be stored for very long without rusting.

Scott realized that Uehling's ideas were superior to his own as soon as he saw the patent sketches. 'Uehling has got what we want,' he declared. Scott's judgement was confirmed when tests at Lucy went extremely well after engineers had created equipment implementing Uehling's design. 'Due principally to the practical ability and honest effort of Mr. Scott,' Uehling wrote, 'the casting machine was, from the start, an unqualified success. The Lucy furnaces became a real Mecca for doubting iron men, as well as interested blast furnace managers. They came and were convinced that the output of the largest blast furnace could be successfully handled over a machine from the furnace to the railroad car without being touched by hand.'

Continuing to reside in Newark, Uehling appointed a representative, J.W. Miller, to negotiate the details of a new agreement with Carnegie. This contract stipulated that Carnegie would pay royalties to Uehling instead of the flat fee that the inventor had originally suggested. Soon, Uehling's device was adopted by other producers in western Pennsylvania and Ohio's Mahoning Valley, who utilized the invention under licence from the American Casting Machine Company, a firm established to control Uehling's patents.

As increasing numbers of northern firms adopted his device, Uehling turned in 1899 to licensing its use abroad. Crossing the Atlantic, Miller organized the English Uehling Company and came back to America with a liberal cash advance from promoters who were interested in exploiting the machine, both in Great Britain and on the European continent.[30]

The manner in which Uehling's machine was adopted abroad once again revealed the impact of differing economic and social circumstances. During the late 1890s, conditions were bad for its reception in Great Britain. It was a time during which British iron and steel manufacturing became what J.C. Carr and W. Taplin have characterized as 'The Static Industry', when the country was 'declining from the mellow splendour of the Diamond Jubilee through self-conscious imperialism and social upheaval into confusion, doubt and war.' American tariff policy had seriously weakened the foundations of the Welsh tinplate industry, causing anxiety among British bankers about supporting innovation in iron and steel generally. Shipbuilding in Great Britain surged for a time, but then declined at the turn of the century, further shrinking markets for iron and steel. The United States had made great inroads on Canadian markets for British ferrous metals, and Belgian and German manufacturers were invading other colonial outlets previously dominated by the mother country. Parliament was indecisive about imposing tariffs to counter a rising tide of ferrous metal imports from America and the European continent. British iron and steel producers could not make up their minds about how to deal with the transition from Bessemer to open-hearth technology. The result of these various circumstances was that the average physical size of a British blast furnace remained small.[31]

Years later, Uehling recalled his disappointment about the lack of response to his pig-casting machine in Great Britain. 'Although England is the mother of the iron and steel industry, the English ironmasters allowed themselves to be outstripped by Germany, as well as the United States,' he wrote. 'The outputs of the English blast furnaces were . . . comparatively small and could be handled by hand without great difficulty. Labor being cheap, the English ironmasters felt they could not justify the considerable outlay necessary to install the pig casting machine. The managers of the English iron works were greatly handicapped not only because their plants, especially their blast furnaces, were antiquated, but because they were built with the idea of permanence. All solid, mostly stone construction, difficult and expensive to modify or enlarge. This was also true of their equipment.'[32]

The situation on the European continent was much different. Partly because their fuel costs were higher than those of British iron and steel manufacturers, Continental industrialists had long been more receptive than their British counterparts to technological innovations aimed at driving down expenses. The advent of the

Thomas–Gilchrist method of producing basic steel in 1879 had stimulated a race among leading continental producers such as Schneider (France), Wendel (German Lorraine) and the Hörder-Verein and Rheinische Stahlwerke (Germany) to fortify their patent positions and embrace new methods. While most British producers temporized and clung to the acid Bessemer process, European firms expanded basic steel production rapidly, and also did more than English counterparts to integrate steel mills and blast furnaces. As a result, furnace sizes in Germany and Austria-Hungary increased dramatically; by the 1890s, the largest British furnaces were modest by comparison. As David Landes indicates, such circumstances promoted particularly rapid mechanization in Germany, where 'continuous conveyors, travelling cranes, and suspended railways powered by electricity' became common in ferrous metals production at the turn of the century.[33]

After being able to interest only one British company in the pig-casting machine, Uehling and Miller turned to central Europe. Because Uehling had been fluent in German since his childhood, he accompanied Miller on a trip to interest German-speaking industrialists in his invention. At Aachen, he and Miller had no problem negotiating a lucrative deal with officials of Aachener Hütten Verein covering that firm's eight existing blast furnaces and any further installations that it might erect in the future. From Aachen, Uehling and Miller went on to Vienna, where the two largest furnace companies in Austria-Hungary had their headquarters. In the Austrian capital, as Uehling later stated, 'the general manager of the Alpina Montan Gesellschaft invited us to a fine dinner at his magnificent residence' and readily signed a contract to utilize the inventor's pig-casting apparatus. Such occurrences typified Uehling's encounters with German-speaking capitalists who were pioneering industrial research laboratories at the time and who impressed him by their willingness to give scientists and engineers due recognition and reward. In his autobiography, Uehling stated that 'Germans were much more ready to recognize the merits of the inventor and more willing to pay a fair royalty than the Americans . . . I found the Germans most liberal in their consideration of the inventor.'[34]

Almost as soon as Uehling's pig-casting apparatus was introduced in America, many competing systems were developed. Only three, however, proved operable. Two of them, embodied in what came to be known as Davis and Heyl-Patterson machines, were acquired by the American Casting Machine Company and consolidated with Uehling's patents. For reasons of space, the details of these systems cannot be analysed here. It is sufficient to say that by the end of his career Uehling could justly claim that his apparatus had 'stood the test of time, and is now in use at practically all blast furnaces'.[35]

Before he died in 1952, Uehling also had the satisfaction of knowing that the Sloss-Sheffield Steel and Iron Company, the successor of the Alabama firm for which he had worked four decades earlier, had adopted the general features of his system in a form using Heyl-Patterson equipment.[36] Market-related circumstances helped make this a sensible decision. The high phosphorus content of Alabama iron ore, which made such ore unsuitable for pig iron intended for conversion into steel, was an advantage with regard to making pig iron for foundry use because it promoted liquidity and ease in moulding. As a result, the American foundry trade had moved steadily into the Southeast since the 1890s, putting a large number of installations within easy reach of Birmingham. Alabama had become the nation's leading producer of soil pipe and pressure pipe, and the South had spawned a variety of plants making stoves, steam engine parts, enamel-plated kitchen and bathroom fixtures, engine blocks, and piston rings. TCI's move into steel manufacture (in 1907, this enterprise was acquired by the United States Steel Corporation) gave such firms as Sloss-Sheffield and the Woodward Iron Company a clear field to exploit growing demand for foundry pig iron. Lack of competition from TCI made possible a higher volume of foundry pig iron production at Sloss-Sheffield than that which had prevailed at the time when Uehling had lived in Birmingham. As demand increased, it was sometimes hard for the company to produce an adequate volume of iron with its existing facilities. At the same time, pig iron prices were highly unstable prior to World War I and fell precipitously during the 1920s. This made rationalization of operations increasingly necessary.[37]

Sloss-Sheffield's leadership had also changed. After World War I, the firm had been acquired by a syndicate headed by Bernard Baruch, a South Carolinian who had been President Woodrow Wilson's chief industrial adviser. Other key directors included Waddill Catchings, a native of Tennessee who had strong Wall Street ties and ultimately became a partner in Goldman, Sachs and Company; and Matthew Scott Sloan, an Alabama-born engineer who had become a prominent electrical utilities executive in New York City. The new owners were closely associated with Allied Chemical and Dye Corporation, whose Semet-Solvay subsidiary controlled patents under which Sloss-Sheffield built a large by-product coke plant at North Birmingham. Like Joseph Bryan and his Richmond associates, the new controlling syndicate had a strongly southern composition. Unlike the old ownership, however, it had a strong scientific orientation and enjoyed much better access to far larger amounts of capital, particularly because leading Wall Street financiers were now well accustomed to investing in industrial securities.[38]

Sloss-Sheffield's new directors were highly committed to innovative production methods. Beginning in 1926, the company's blast furnaces were rebuilt by James Pickering Dovel, the most creative engineer to work for the firm since Uehling had resigned in 1895. The new furnaces could produce up to 450 tons of iron, a volume difficult to handle with traditional sand-casting methods. Dovel introduced many new design features that were aimed primarily at better handling of southern raw materials. Throughout its history, the company had experienced problems with lean ore and low-quality coal containing high percentages of chemical impurities that adversely affected temperature control, produced excessive amounts of slag, left heavy residues on furnace walls, and resulted in severe scaffolding. To cope with such difficulties, Dovel installed better equipment for cleaning furnace gases; redesigned the bosh for more effective cooling; reconfigured interior furnace wall profiles to enhance air circulation and reduce surface friction; and replaced earlier labour-intensive charging methods with modern apparatus that included a new stock trestle and skip hoists.

It was in connection with this larger package of design modifications that Sloss-Sheffield's owners, for the first time in the company's history, adopted automatic pig-casting. Because of scepticism within the foundry trade regarding the abandonment of sand-casting methods, Sloss-Sheffield created a trade journal, *Pig Iron Rough Notes*, for circulation among foundry owners. Bristling with articles by such scientists as Richard Moldenke, America's leading academic authority on foundry practice, the magazine vigorously preached the advantages of machine-cast pig iron.[39]

In a thought-provoking essay, Gary Kulik has argued that Sloss-Sheffield's belated adoption of automatic pig-casting was due to a migration of African-Americans from the South to northern cities that had begun at about the time of World War I, making it more difficult to secure low-paid black workers on whom the company had previously relied. This claim, however, is not convincing. Black migration during and after World War I did not simply flow from South to North; it also involved an exodus of blacks from rural districts to large cities within the South itself. This happened in Alabama; Birmingham's black population actually inceased by about 22,000 during the 1920s. Although there is some evidence that Sloss-Sheffield suffered from temporary shortages of workers during and immediately after the war, it appears unlikely that local industrialists had much cause to worry about labour supply thereafter. As had always been true, labour costs remained less crucial to Sloss-Sheffield and other southern furnace companies than costs associated with the handling and processing of raw materials.[40]

It is conceivable that ongoing demographic shifts in the 1920s created uncertainties about future labour supply in the minds of Sloss-Sheffield's owners. It is also possible that experienced black workers were departing for the North while unseasoned recruits were coming to Birmingham from the countryside. But such things are at best conjectural in the absence of further evidence. The chief factors that led Sloss-Sheffield to adopt mechanical pig-casting after having long failed to embrace it are quite clear. These factors included increasing markets for foundry pig iron; higher production volumes that were difficult to manage with hand labour; declining prices; the coming to power of new owners with modern attitudes; and a desire to achieve a greater degree of scientific control over operations. The degree to which cheap labour was readily available figured only marginally, at most, in the decision-making process that led to the adoption of mechanical casting.

None the less, the circumstances surrounding Sloss-Sheffield's long-delayed adoption of mechanical casting indicate clearly that changes in the surrounding milieu had once again affected technological transfer in the 1920s and early 1930s, just as had been the case at the time when Uehling invented his pig-casting machine a few decades earlier. After being conceived in Birmingham in the 1890s and winning acceptance outside the region, automatic pig-casting had at length returned to its southern place of origin. Throughout the history of the new technique, the cultural context within which it had spread from one place to another had been all-important with regard to the reception it had received. In this respect, the ferrous metals industry was no different from many others, conforming to patterns of development that have already been explored by such historians as Edward N. Constant, Thomas Parke Hughes and Merritt Roe Smith. As in the case of the jet engine, electrical power systems, interchangeable parts, and other innovations too numerous to mention, the history of Uehling's ingenious machine demonstrates the interconnectedness of technology and other aspects of human culture.

Notes and References

*This essay is derived from my forthcoming book *Sloss Furnaces and the Rise of the Birmingham District: An Industrial Epic*, University of Alabama Press. Research for this project was supported in part by Summer Stipend No. FT-33831 (1990) from the National Endowment for the Humanities, Washington, DC, and by a series of Mellon Fellowships from the Virginia Historical Society, Richmond, Virginia. Grateful acknowledgement is also made to Dr Paul F. Parks, Provost and Vice-President for Academic Affairs, Auburn University; the City of Birmingham, Albama; the Hagley Museum and Library, Wilmington, Delaware; and Sloss Furnaces National Historic Landmark, Birmingham, Alabama for additional grants-in-aid.

1. On various features of this revolution, see J.E. Johnson, Jr, *Blast-Furnace Construction in America* (New York, 1917), 15, 49, 61, 66–7, 102, 288–9, 328, 333.

2. Edward A. Uehling, 'Autobiography of Edward A. Uehling, M.E., Dr of Engineering, 1849–1939', unpublished manuscript, 1–36, hereinafter cited as Uehling Autobiography; letter of recommendation for Uehling by Henry Morton, 14 August 1878, and letter from Morton to Uehling, 25 March 1890, in collection hereinafter cited as Morton–Uehling Correspondence. All of these materials were donated to Sloss Furnace National Historic Landmark, Birmingham, Alabama by Uehling's grandson, Dr Edward R. Uehling, and are used with permission of the latter. For a brief account of Morton's life and career, see obituary in *New York Times*, 10 May 1902.

3. For a brief perspective on the developments discussed here and in the paragraphs that follow, see W. David Lewis, 'Joseph Bryan and the Virginia Connection in the Industrial Development of Northern Alabama', *Virginia Magazine of History and Biography*, 1990, 98: 613–32. A more extended account will be found in Chs 6–7 of my *Sloss Furnaces and the Rise of the Birmingham*

District: An Industrial Epic, University of Alabama Press (forthcoming). The standard work on the history of ferrous metals manufacture in Alabama is Ethel M. Armes, *The Story of Coal and Iron in Alabama* (Birmingham, 1910); for details on the Sloss Iron and Steel Company and its successor firm, Sloss-Sheffield Steel and Iron Company, see pp. 347–53, 451–8. In general, however, Armes's pioneering work is badly out of date and contains numerous errors.

4. Letters from Morton to Uehling, 14 and 28 February and 8 and 25 March 1890, Morton–Uehling Correspondence. For a brief account of Dashiell's career, see *National Cyclopedia of American Biography*, 34: 96–7.

5. Uehling Autobiography, *op. cit.* (2), 36–7; entries of 27–30 September, 1890, Record Books of Henry Morton, S.C. Williams Library, Stevens Institute of Technology, Hoboken N.J., cited with permission of Stevens Institute of Technology.

6. Edward A. Uehling, 'Pig-Iron Casting and Conveying Machinery', *Cassier's Magazine*, 1903, 24: 115–18. See also Gary B. Kulik, 'The Sloss Furnace Company, 1881–1931: Technological Change and Labor Supply in the Southern Pig Iron Industry', unpublished research report prepared for Historic American Engineering Record, Washington, DC, 1976, 7–8, and Kulik's 'Black Workers and Technological Change in the Birmingham Iron Industry, 1881–1931' in Merl E. Reed, Leslie S. Hough and Gary M. Fink, eds, *Southern Workers and Their Unions, 1880–1975* (Westport, Conn., 1981), 26–7.

7. Specification for US Patent No. 548,146, 'Apparatus for and Method of Casting and Conveying Metals', 15 October 1895; Uehling, 'Pig-Iron Casting and Conveying Machinery', *op. cit.* (6), 117–19. On the same point, see Albert L. Colby, 'The Advantages of Machine-Cast Pig-Iron', *Cassier's Magazine*, 1903, 24: 134–8. On the various grades of iron that were produced by southern blast furnaces at this time, see weekly market reports on Cincinnati, Ohio and Louisville, Kentucky during the 1880s in a leading trade publication of the American ferrous metals industry, *Iron Age*.

8. See, particularly, testimony of James W. Sloss in United States Senate, *Report of the Committee of the Senate Upon the Relations Between Labor and Capital, and Testimony Taken by the Committee*, 5 vols (Washington, DC, Government Printing Office, 1885), 4: 281–91.

9. 'Religious Observations and Reflections', Uehling Autobiography, *op. cit.* (2), 121–33; Uehling, 'Pig-Iron Casting and Conveying Machinery', *op. cit.* (6), 118.

10. Uehling Patent Specification, 1–4 with accompanying illustrations. I am indebted to Melvin C. Smith, a graduate student in the Department of History at Auburn University, for assistance in studying the details of Uehling's process and the functioning of rival systems, mentioned later in this article, that were appearing in such places as western Pennsylvania at about the same time.

11. Uehling Autobiography, *op. cit.* (2), 40, 44–5; letter from Morton to Uehling, 17 March 1894, Morton–Uehling Correspondence; Specification for US Patent No. 554, 323, 'Pyrometer', 11 February 1896; *Iron Trade Review*, 6 May and 3 June 1897.

12. On Bryan's admiration for Uehling and approbation by Bryan and other directors for the engineer's work, see letters from Bryan to Morton, 2 June and 5 July 1890 and from Bryan to Seddon, 4 February 1895, Joseph Bryan Letterbooks, Virginia Historical Society, Richmond, Virginia, cited with the society's permission. In the same connection, see Uehling Autobiography, *op. cit.* (2), 38–40.

13. F.J. Crolius, *op. cit.* (2), 'Straight Line Production', *Coal Industry*, June 1924, 224. Among numerous analyses of the circumstances affecting ironmaking in the area, see particularly R.H. Ledbetter, 'Blast Furnace Practice in Birmingham District', *Iron Age*, 30 October 1924, 1128–30.

14. Jack Roland Bergstresser, Sr, 'A Brief History of the Birmingham District', in *The Birmingham District: An Assessment*, report prepared for National Park Service by Birmingham Historical Society, Birmingham, Alabama, June, 1992, 34–5. Estimates of the cost per ton of pig iron produced in the District at the time Uehling came there ranged from as low as $9 to a maximum of $12; see particularly *Iron Age*, 22 December 1887, 29; *ibid.*, 24 May 1888, 857; and Victor S. Clark, *History of Manufactures in the United States* (New York, 1929), 2: 287.

15. I.W. Avery, 'Southern Men and Southern Money Developing the South', *Manufacturers' Record*, 2 August 1884; Paul H. Buck, *The Road to Reunion 1865–1900* (Boston, 1937), 152–7; Gavin Wright, *Old South, New South: Revolutions in the Southern Economy since the Civil War* (New York, 1986), 13.

16. For the best accounts of northern investment in southern railroading after the Civil War, see Maury Klein, *The Great Richmond Terminal: A Study in Businessmen and Business Strategy* (Charlottesville, Va., 1970), and John F. Stover, *The Railroads of the South, 1865–1900: A Study in Finance and Control* (Chapel Hill, N.C., 1955). On the reluctance of northern investment houses to risk large amounts of capital in manufacturing until late in the nineteenth century, see particularly Nathan Rosenberg and L.E. Birdzell, Jr, *How the West Grew Rich: The Economic Transformation of the Industrial World* (New York, 1986), 220–8.

17. Lewis, 'Joseph Bryan and the Virginia Connection', *op. cit.* (3). For further insight into Bryan's 'southern economic nationalism' and overall business outlook, see James M. Lindgren, 'First and Foremost a Virginian: Joseph Bryan and the New South Economy', *Virginia Magazine of History and Biography*, April 1988, 157–80.

18. For a useful survey of problems that inhibited steelmaking in Alabama in the late nineteenth century, see Justin Fuller, 'From Iron to Steel: Alabama's Industrial Evolution', *Alabama Review*, April 1964, 137–48.

19. For the best available account of TCI's successful effort to produce basic steel, see Justin Fuller, 'History of the Tennessee Coal, Iron, and Railroad Company, 1852–1907', Ph.D. dissertation, University of North Carolina at Chapel Hill, 1966, 250–67. On the marketing strategies pursued by ironmakers in the District in the late nineteenth century, see particularly W. David Lewis, 'The Invasion of Northern Markets by Southern Iron in a Decade of Boom and Bust: Sectional Competition in the 1880s', *Essays in Economic and Business History*, 1990, 8: 257–69.

20. A searching account of these and other developments is provided in Jack Roland Bergstresser, Sr, 'Raw Material Constraints and Technological Options in the Mines and Furnaces of the Birmingham Distrct: 1876–1930', Ph.D. dissertation, Auburn University, 1993.

21. James Saxon Childers, *Erskine Ramsay: His Life and Achievements* (New York, 1942), 197–233, 429–533 and *passim*.

22. J. Mills Thornton, *Politics and Power in a Slave Society: Alabama, 1800–1860* (Baton Rouge, La., 1978), 7–11, 194, 272–3, 291.

23. Lindgren, 'First and Foremost a Virginian', *op. cit.* (17), 161.

24. Robert S. Starobin, *Industrial Slavery in the Old South* (New York, 1970), *passim*; Woodward Iron Company, *Alabama Blast Furnaces* (Woodward, Alabama, 1940), 17, 46, 51, 74, 84, 87, 105, 120, 122, 126, 136; Carroll W.

Pursell, *Early Stationary Steam Engines in America: A Study in the Migration of a Technology* (Washington, DC, 1969), 75; Charles B. Dew, 'Southern Industry in the Civil War Era: Joseph Reid Anderson and the Tredegar Iron Works, 1859–1867', Ph.D. dissertation, Johns Hopkins University, 1964, 346–85; Charles B. Dew, *Ironmaker to the Confederacy: Joseph R. Anderson and the Tredegar Iron Works* (New Haven, Conn., 1966), 22–32; Richard Wade, *Slavery in the Cities: The South, 1820–1860* (New York, 1964), 34–5, 243–81; Claudia Dale Goldin, *Urban Slavery in the American South, 1820–1860: A Quantitative History* (Chicago, 1976), *passim*; Robert W. Fogel and Stanley L. Engerman, *Time on the Cross: The Economics of American Negro Slavery* (Boston, 1984), *passim*; Carroll W. Pursell, 'Cyrus Hall McCormick and the Mechanization of Agriculture', in Carroll W. Pursell, ed., *Technology in America: A History of Individuals and Ideas*, 2nd edn (Cambridge, Mass., 1990), 71–9.

25. John T. Milner, *Report of the Chief Engineer to the President and Board of Directors of the South and North Alabama Railroad Company* (Montgomery, Alabama, 1859), 13–14; John T. Milner, *Alabama: As It Was, As It Is, and As It Will Be* (Montgomery, 1876), 146–61; Jack L. Lerner, 'A Monument to Shame: The Convict Lease System in Alabama', MA thesis, Samford University, 1969, *passim*. For vignettes of black subcontractors at the turn of the twentieth century, see *Bessemer Weekly*, 18 May 1901. On the centrality in post-Civil War southern development of a white supremacist belief that forced labour was the only suitable system for black workers, see also Eric Foner, *Reconstruction: America's Unfinished Revolution 1863–1877* (New York, 1988), 128–42, 198–9, 213–14, 372–3, 405–6, 419, 549, 593, 597.

26. On the steady growth of Birmingham's black population, see *The Mineral Wealth of Alabama and Birmingham Illustrated* (Birmingham, 1886), 109; and Robert J. Norrell and Otis Dismuke, *The Other Side: The Story of Birmingham's Black Community* (Birmingham, n.d.), 1–2. On the city's reputation as the 'Johannesburg of America', see Horace Huntley, 'Iron Ore Miners and Mine Mills in Alabama: 1933–1952', Ph.D. dissertation, University of Pittsburgh, 1977, 16. On the similarity of conditions in Birmingham and Johannesburg, compare photographs in Leah Rawls Atkins, *The Valley and the Hills: An Illustrated History of Birmingham and Jefferson County* (Woodland Hills, Calif., 1981) and Wayne Flynt, *Mine, Mill and Microchip: A Chronicle of Alabama Enterprise* (Northridge, Calif., 1987), with those in Peter Kallway and Patrick Pearson, *Johannesburg: Images and Continuities* (Braamfontein, South Africa, 1986). For a comparative discussion of conditions in Alabama's rural districts and industrial centres, see Jingsheng Dong, 'The Social and Working Lives of Alabama Coal Miners and Their Families, 1893–1925', MA thesis, Auburn University, 1988, 39–40 and *passim*.

27. For a brief overall discussion of these trends, see W. David Lewis, *Iron and Steel in America* (Wilmington, Delaware, 1976), 35–54. Among other sources, see particularly James H. Bridge, *The Inside History of the Carnegie Steel Company: A Romance of Millions* (New York, 1903), *passim*; Alfred D. Chandler, *The Visible Hand: The Managerial Revolution in American Business* (Cambridge, Mass., 1977), 259–69, 360–2; William T. Hogan, *Economic History of the Iron and Steel Industry in the United States*, Vol. 1 (Lexington, Mass., 1971), *passim*; Thomas Jay Misa, 'Science, Technology and Industrial Structure: Steelmaking in America, 1870–1925', Ph.D. dissertation, University of Pennsylvania, 1987, 1–59; Peter Temin, *Iron and Steel in Nineteenth Century America: An Economic Inquiry* (Cambridge, Mass., 1964), 125–230; David A. Walker, *Iron Frontier: The Discovery and Development of Minnesota's Three Ranges* (St. Paul, Minn., 1979), *passim*; Joseph Frazier Wall, *Andrew Carnegie*, 2nd edn (Pittsburgh,

1989), 307–60, 471–672; and Kenneth Warren, *The American Steel Industry, 1850–1970: A Geographical Interpretation* (London, 1973), 88–159.

28. Uehling Autobiography, *op. cit.* (2), 47–8. On burdens of 600 tons being reached by some installations in the Pittsburgh District by the mid-1890s, see Johnson, *Blast-Furnace Construction in America, op. cit.* (1), 15.

29. David Brody, *Steelworkers in America: The Nonunion Era* (New York, 1960), 27–49. As Brody indicates, wage rates were 'lowest in the South, highest in Pittsburgh and Chicago'.

30. Uehling Autobiography, *op. cit.* (2), 47–50.

31. J.C. Carr and W. Taplin, *History of the British Steel Industry* (Cambridge, Mass., 1962), 183–96.

32. Uehling Autobiography, *op. cit.* (2), 52–3.

33. David Landes, *The Unbound Prometheus: Technological Change and Industrial Development in Western Europe from 1750 to the Present* (Cambridge, 1969), 249–69.

34. Uehling Autobiography, *op. cit.* (2), 52–3, 58–61.

35. Uehling, 'Pig-Iron Casting and Conveying Machinery', *op. cit.* (6), 125–33.

36. For details of the combined system, see Johnson, *Blast-Furnace Construction in America, op. cit.* (1), 340–4.

37. W. David Lewis, 'Sloss Furnaces and the Southern Foundry Trade: A Case Study of Industrialization in the New South', in Howard L. Hartman, ed., *Proceedings of the 150th Anniversary Symposium on Technology and Society: Southern Technology, Past, Present, and Future* (Tuscaloosa, Ala., 1988), 58–80.

38. The discussion provided here is based on a much longer version in my book, *Sloss Furnaces and the Rise of the Birmingham District* (Birmingham, Ala., forthcoming), Ch. 14, with documentation too extensive to cite fully in this essay. For biographical data on Sloss-Sheffield's new directors, see article on Baruch by Robert D. Cuff in *Dictionary of American Biography*, Supplement 7 (1981), 34–7; article on Catchings by Broadus Mitchell in *Dictionary of American Biography*, Supplement 8 (1988), 78–9; and obituary of Sloan in *New York Times*, 15 June 1945.

39. 'Rebuilding Alabama Blast Furnaces', *Iron Age*, 24 June 1926; 'Birmingham's Most Modern Iron Maker' and 'Sloss New No. 2 Furnace', *Pig Iron Rough Notes* (hereinafter cited as PIRN, August–September 1927; 'Dovel Type Blast Furnace Put on Test' and 'Improvement of Existing Blast Furnaces', *Blast Furnace and Steel Plant*, December 1928 and August 1930, clippings at Slos Furnaces National Historic Landmark, Birmingham, Alabama; 'Sloss Ladle–Mixed Pig Iron', PIRN, Fiftieth Anniversary Issue, 1932, 24–6. On the early history of the new trade journal, see Russell Hunt, 'Rough Notes Begins Ninth Year', PIRN, Autumn 1934.

40. Kulik, 'Black Workers and Technological Change', *op. cit.* (6), 33–6; James L. Sledge III, 'The Great Migration, Alabama Blacks, and the Birmingham District: An Overview', unpublished research paper, Auburn University, 1993; John Jermain Porter, 'The Manufacture of Pig Iron', in A.O. Backert, ed., *The ABC of Iron and Steel* (Cleveland, Ohio, 1915), 86. For an extended argument that market-related circumstances, particularly declining pig iron prices, had more to do than labour supply with the technological changes made by Sloss-Sheffield in the late 1920s and early 1930s, see Ch. 14 of my forthcoming book *Sloss Furnaces* (*op. cit.* (38)).

The Toledo Water-Clocks of *c.* 1075

DONALD HILL†

This paper concerns the two large clepsydras which were erected on the banks of the river Tagus at Toledo in the second half of the eleventh century. The principal object of the study is to attempt to reconstruct on paper the clocks very much as they may have existed at the time, together with an explanation of how they worked. It will be necessary also to outline briefly the historical background of water-clocks, especially in Spain, and to discuss some difficulties in the interpretation of the text attributed to Archimedes.

The history of complex water-clocks began in Hellenistic times, one of the most important texts being a treatise attributed to Archimedes. Although the Arabic text in which it survives contains Byzantine and Islamic additions, the very ingenious water machinery was probably the invention of Archimedes himself.[1] The most important work on water-clocks in eastern Islam is contained in the great machine book of al-Jazarī, completed in 1206 in Diyār Bakr. The first and longest section in the book describes the construction and operation of six water-clocks and four candle-clocks. In the other five sections, describing various types of automata and a few utilitarian machines, many of the components, mechanisms and techniques are analagous to those used in the construction of the clocks.[2]

The earliest reference to a water-clock in Islamic Spain occurs in a text by ʿAbbās b. Firnās (d. 887), poet and engineer to the emirs ʿAbd al-Raḥmān II and Muḥammad I.[3] In the following century there is a treatise, existing only in a badly defaced manuscript, attributed to a certain al-Murādī. The manuscript was discovered only in the 1970s. Because of its condition it is doubtful whether it will ever be possible for the water-clocks and automata described in it to be fully elucidated. Some significant aspects of their construction can, however, be ascertained. The hydraulics of these machines, when compared to al-Jazarī's work, are quite crude, but the transmission systems are important in the history of horology. They incorporate gear-trains embodying segmental and possibly epicyclic gears. The treatise is also significant in that it describes the use of mercury inside tilting balanced pipes.[4]

The work entitled *Libros del saber de astronomía* was commissioned by Alfonso X of Castile with the express purpose of making translations and paraphrases of Arabic works into Castilian so that Islamic knowledge could be made available to Christians. It was completed in 1277. The work includes five timepieces, described by Isaac ibn Sid, of which the most interesting is a compartmented, cylindrical mercury clock. It consisted of a large drum made of walnut or jujube wood tightly assembled and sealed with wax or resin. The interior of the drum was divided into twelve compartments, with small holes between the compartments through which mercury flowed. Enough mercury was enclosed to fill just half the compartments. The drum was mounted on the same axle as a large wheel powered by a weight-drive wound around the wheel. Also on the axle was a pinion with six teeth that meshed with thirty-six oaken teeth on the rim of an astrolabe dial. The mercury drum and the pinion made a complete revolution every four hours and the astrolabe dial made a complete revolution in twenty-four hours. This clock was said by Isaac to have been built two hundred years earlier, which indicates that the Muslims may have known the principle of weight-driven clocks two centuries before it was appreciated in Europe.[5]

In attempting a conjectural but, it is hoped, credible, reconstruction of the Toledo clock, only mechanisms that are known to have been used by Islamic constructors of water-clocks and similar devices will be incorporated in the machine. These will be discussed individually when the assembly and operation of the clock is described. There is, however, one matter relevant to time measurement in all pre-mechanical clocks to be mentioned — namely, the question of temporal and equal hours. Temporal or solar hours are the units that are obtained by dividing the hours of daylight — or darkness — by twelve, giving 'hours' that vary in length from day to day throughout the year. Some sort of flow regulator was therefore necessary to achieve this daily variation. Equal hours are simply our units of 60 minutes each.

It is commonly asserted by European historians of technology that equal hours were never used before the invention of the Western mechanical clock. It is on this account that dates in the early years of the fourteenth century are given for the first appearance of clocks working on equal hours.[6] It is simply not the case, however, that Islamic water-clocks never used equal hours. In part of a book written in 1121/2 the famous physicist al-Khāzinī described a steelyard clepsydra that worked equal hours,[7] and in the previous century, as we have seen, the eleventh-century Islamic original of the mercury clock described in the *Libros del saber* worked on equal hours. Of the 10 clocks described by al-Jazarī in 1206, but constructed during the previous two decades, two worked on temporal hours, the remainder on equal hours.

The construction of the Toledo clock is usually attributed to the astronomer al-Zarqāllu (d. 1100). It is mentioned by several medieval Arabic writers. The text usually cited is that of Abu'l-ʿAbbās al-Maqqārī (d. 1632), in *Nafḥ al-ṭīb* (latest edition Beirut, 1968), pp. 206–7. This is now given in English translation:

Among the wonders of al-Andalus [were] the two basins in Toledo constructed by ᶜAbd al-Raḥmān when he heard of the talisman which was in the city of Arīn in the land of Hind.[8] It was mentioned by al-Masᶜūdi (d. 956). It rotated with its pointer (lit. finger) from dawn until sunset. [ᶜAbd al-Raḥmān] made these two basins outside Toledo in a hollow house in a hollow of the great river [i.e. the Tagus] at the place known as the 'gate of the tanners'. Their wonder lay [in the fact that] they filled and emptied with the waxing and waning of the moon, namely, at the appearance of the first crescent a little water emerged into them; at daybreak a quarter of a seventh [of their capacity] of water was in them; at the end of the day there was completed in them half a seventh. And so on, night and day half a seventh until seven days of the month had elapsed and seven nights, at which point there was in them half of their capacity. This increase continued — half a seventh in the day and the night until at full moon they were completely full. When it was the fifteenth night and the moon began to wane they decreased with the waning of the moon, every day and night half a seventh until the moon had completed twenty-one days, and the contents had decreased by one half; and so it continued, decreasing in every day and night half a seventh. On the twenty-ninth of the month there remained no water in them.

If someone took it upon himself to fill them, and brought water to them, they would absorb that until there was nothing left in them except what should be in them at that time. And if someone took it upon himself when they were filled to empty them and leave nothing in them, when he raised his hand from them water would run into them which would fill them to the correct level for that instant.

They were more marvellous than the talisman of India because that was in the equatorial zone where night does not exceed the day, whereas these were not like that, not being in an equatorial place.

They remained in one house until the Christians (may Allah ruin them) possessed Toledo [in 1085] and Alfonso [VII of Castile] wished to know how they worked and ordered one of them to be dismantled to see how the water and how the movement came to them. So it was dismantled and its movement was spoiled. That was in the year 538 (1133/4). It is said that the cause of their destruction was Ḥunayn the Jew who brought the doves of al-Andalus to Toledo in a single day in the year 527. It was he who told Alfonso that his son would enter Cordoba and take possession of it. He wanted to discover the mechanism of the basins and he said, 'Oh king, I will dismantle them and reassemble them better than they were [before]. I will make them fill during the day and evacuate in the night'. And when they were dismantled he could not put them together again. It is said that he dismantled one of them to obtain (lit. steal) its method of construction, and it was spoiled. The other one went on working. And Allah knows the truth of the matter.

Several European scholars have discussed the Toledo clock. The passage in al-Maqqārī was translated into English by Pascual de Gayangos in his version of *Nafḥ al-ṭīb* entitled *The History of the Muhammadan Dynasties in Spain* (2 vols, London, 1840–3, Vol. 1, 81). The chief defect of Gayangos's translation is his interpolation of words and phrases which are not is the Arabic original, without any indication that this has been done. For instance, he adds, after the words 'Abdi-r-rahman' (*sic*) the gloss 'known by the name of Az-zarkal', words which are not in the Arabic text. This and other similar interpolations detract from the value of his translation. In the *Zeitschrift für Mathematik und Physik, Historisch-Literarische Abteilung*, 39, 1984, 41–55, Armin Wittstein reproduced the Arabic text, Gayangos's translation and his own German translation. The translation is adequate and Wittstein also provided annotations. The passage was also translated into Spanish by J.M. Millás Vallicrosa, *Estudios sobre Azarquel*, Madrid, 1950, 6–9. Other brief references can be found in *De Abd al-Rahman I a Isabel II* by Juan Vernet (Barcelona, 1989, 297–300) and in *Islamic and Christian Spain* by T.F. Glick (Princeton, 1979, 244–5).

It has to be said that none of these writers attempts to assess the practicability of these clepsydras. Indeed, Wittstein suggests that they may have existed only in tradition. He also doubts whether they could have been constructed by a competent astronomer such as al-Zarqāllu because they have a total time cycle of 28 days, whereas the synodical month has an average value of 29.53 mean solar days.

Before the reconstruction of the clock is described it will therefore be necessary to confront some of the difficulties and anomalies in the text. Compared with the very detailed and scrupulous descriptions and specifications given by al-Jazarī, for example, this text is very short and gives no constructional details. Moreover, it would seem to have been written by a layman who had no knowledge of horology or engineering, and perhaps only a sketchy acquaintance with astronomy.

The first problem is the aforementioned use of a unit of 28 days for the synodical month (the interval between two 'new moons'). It is considered likely that the actual period recorded was close to the true figure of 29.53 days, or 14.76 days for half a month, and that manual adjustments were made periodically to the water levels to ensure that the clock coincided with the months and years.

Considering now the question of the corrections necessary if water were added during evacuation or extracted when the vessels were full, and taking the second case first: this could be achieved quite simply by using a mechanism described by the Banū Mūsà brothers (Baghdad, mid-ninth century) in several of their devices. This was very similar in operation to a modern ball valve, whereby a float caused a valve to open and replenish a tank if water were extracted from it.[9] There seems to be no way, however, with the techniques available at the time, by which an automatic mechanism could have worked in the first case. If there were, for example, some kind of system using a float, whereby pipes or siphons were opened in the side of the vessel as the water level descended, then if extra water were poured in the float would rise and close these outlets again. Manual operation was the most likely method for adjusting the water level. To simplify the drawings a manual system is suggested for the second case also.

We now have to consider the author's remark that these clepsydras were more marvellous than the 'talisman' of Arīn because that was in the equatorial zone whereas the Toledo clepsydras were not. This statement is nonsense even if the Toledo and Arīn clocks were designed to record temporal hours. Arīn — 'the centre of the world' — is on the equator, but, even so, near the solstices a flow regulator would be required to produce a daily variation in temporal hours. But was it practicable to record temporal hours with the Toledo clepsydras? In clocks such as the first to be described by al-Jazarī, the working face was 1.35 m wide. This contained an array of time-recording automata in a horizontal line. At the winter solstice the day at Diyar Bakr was about 9.5 hours long, each temporal hour being therefore $9.5 \times 60/12 = 47.5$ minutes. In this time the pointer on the face of the dial had to traverse a distance of $1.35/12 = 11.25$ cm, giving plenty of room for the operation of the automata.[10]

Consider now the Toledo clock. Assume daylight in Toledo at the summer solstice to be 15 hours long and at the winter solstice 9 hours, and further assume that the clepsydras were 5 metres high. Half a synodical month is 14.76 days. Then during the filling of the basins the daily rise in water level would be $5/14.76 = 33.9$ cm, or an equal hourly rate of 1.41 cm. The rise in a temporal hour at the summer solstice would be $33.9/(2 \times 15) = 1.13$ cm, and at the winter solstice $33.9/(2 \times 9) = 1.88$ cm. No one, surely, would suggest that such small differences in level, especially when there were apparently no automata, could be used to record temporal hours. The movement up or down of the water level would have been barely perceptible. It seems likely that the transmitter of the report knew little of horology and believed that all clocks worked on temporal hours. It is perhaps significant that the report mentions the movements in the night and the day separately only once. Thereafter the author speaks of changes in level of half a seventh of the capacity of the vessels in a 24-hour period.

So far the reasons given for the necessity for manual intervention while the clock was working are concerned with timekeeping, in particular that the clock recorded true synodical months and that the addition or extraction of water was compensated for. There is, however, a more important reason for assuming that the clock was not wholly self-operating (and the author never says explicitly that it was), and that is the requirement for maintenance. The water cannot have been clean and free from impurities. Although a settling tank has been included in the reconstruction in the supply system (see Figure 1) this would not have been sufficient to clean the water thoroughly. To ensure that orifices remained free from obstructions, constant attention would have been necessary. This point is emphasized in the description of the water-clock in the *Libros del saber*, where a section is devoted to the preparation of the water to ensure an uninterrupted flow through the narrow orifice. The water was placed in a clean vessel and 6 ounces of powdered alum was added to it to precipitate impurities. The water above the precipitate was extracted and poured into another clean vessel. It was then filtered through cloth. Finally it was stored in a covered container until it was required.[11] It is unlikely that such elaborate precautions would have been possible in the Toledo clepsydras, given the large quantity of water required daily. The attendance of a full-time maintenance crew would seem to have been an absolute necessity and the plant must have been shut down regularly in order for thorough maintenance to be carried out. This would have included the cleaning of tanks and pipes and the changing of orifices, since these would have been enlarged by the erosion of the water.

RECONSTRUCTION OF THE CLOCK

Although Figures 1 and 2 are felt to be credible representations of the original construction they are of course conceptual and are intended to show how the clock could have been designed and assembled. Some features that must have been provided in the original machine have been omitted. The most important of these omissions are the supports for the upper tanks, and the ladders and walkways that would have been needed by the supervisors and workmen. Only one clepsydra and its associated hydraulic machinery is shown. An exactly similar arrangement would have been installed immediately adjacent to the one shown.

The text is not clear about the housing of the plant, but it is not necessary to visualize this with any precision. The supply could either have been gravity-fed or conducted to the required level by a noria.

Three large tanks — A, B and C — formed the main part of the clock. Tank C was the clepsydra itself, the remainder of the machinery being shut off from public view by a partition wall that reached from floor to roof. Tanks B and C combined were equal in capacity to tank A, whose function was simply to produce the alternation at half-monthly intervals. At the bottom, tanks B and C were connected by a pipe. (Figure 1).

On a fulcrum on the wall between tanks A and B is a balanced pipe having a funnel on its upper side to receive the water supply and being provided with two short pipes fitted with calibrated orifices-

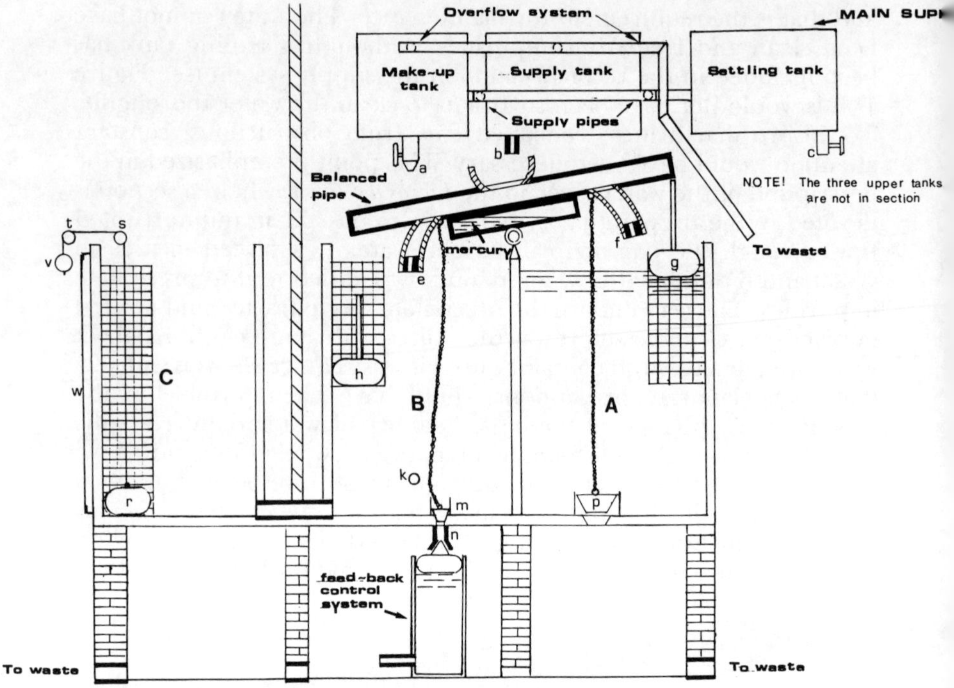

Figure 1 Toledo water-clock: sectional elevation on centre line.

orifice *e* discharged into tank B and orifice *f* into tank A. Soldered to the underside of the balanced pipe was a short closed pipe containing about half of its capacity of mercury. The whole assembly was actually balanced on this short pipe.

There were two rings soldered to the underside of the balanced pipe. One was connected by a chain to a conical valve *p* in the bottom of tank A, the other by a chain to a smaller conical valve *m* in the bottom of tank B. Both valves were restrained by short cylinders made of wire mesh so that they fell into their seats when the tension on the chains was released.

Beneath the floor of the three tanks were three chambers. When valve *p* was opened the contents of tank A ran into the chamber below and then through a pipe to waste. Under valve *m* was a short length of pipe *n* leading to a feed-back control system which ensured that tank B evacuated in a given time (see below). The outlet from this system discharged through a pipe into the third chamber, then through another pipe to waste.

At the sides of tanks A and B were two floats *g* and *h*, each carrying a vertical rod and each restrained by a 'cage' made of wire mesh. In tank B, towards the bottom, was a tap *k*.

In tank C there was a float *r*, also restrained by a wire mesh cage. A

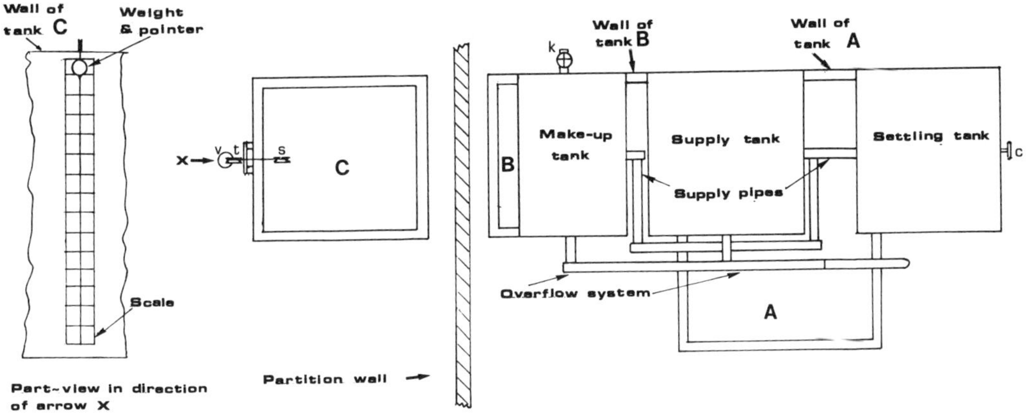

Figure 2 Toledo water-clock: plan to show layout of tanks and pipes only.

string from the top of this float led over the two pulleys *s* and *t* and its end was tied to weight *v* which had a pointer on its underside. This pointer moved over a scale *w* divided into days and hours.

Above the tanks, A, B and C was the supply system (Figures 1 and 2). Water from the mains was first led into a settling tank in which silt and other foreign matter would sink to the bottom.[12] A tap *c* allowed this tank to be emptied before it was periodically cleaned. At about one-third of its height from the bottom, pipes led into the supply tank and the make-up tank. A calibrated orifice *b* dripped water from the bottom of the supply tank into the funnel of the balanced pipe. A tap *a* could discharge water from the make-up tank into tank B. Overflow pipes from the supply tank and the make-up tank were connected to a header pipe which conducted the overflow water to waste.

OPERATION

The system is illustrated in Figure 1 at the moment when the rod on float *g* has just struck the balanced pipe, tilting it towards tank B, the mercury ensuring that it remains in this position. At the same instant valve *p* in the floor of tank A opens and tank A begins to evacuate — the rate of this evacuation is unimportant. Tanks B and C begin to fill through orifice *b*, the balanced pipe, and orifice *e*, the filling being complete after a period of 14.76 days. The weight *v* and its pointer move over scale *w*. After this period the rod on float *h* strikes the balanced pipe, tilting it towards tank A. Valve *p* closes, valve *m* opens and tanks B and C begin to discharge through the feed-back control system. This system was almost certainly invented by Archimedes and was used by both Riḍwān and al-Jazarī.[13] Water runs through pipe *n* whose end is formed into the seat of a conical valve. The plug of this valve is on a float in the float-chamber which is beneath the floor of tank B. The float-chamber is provided with an outlet pipe. Water runs from pipe *n* into the float-chamber, whose valve therefore closes momentarily; discharge now occurs from the outlet pipe, the valve opens momentarily, and so on. A virtually constant head is maintained in the float-chamber and the rate of discharge is therefore constant. Provided that the pipe *n* and the valve in the float-chamber are carefully calibrated, the discharge from tanks B and C can be made to occur at a constant rate that empties the tanks in 14.76 days. Clearly, the orifices *b*, *e* and *f* must also be carefully calibrated to produce the correct rates of flow.[14]

During the emptying of tanks B and C, tank A has been filling; being equal in capacity to tanks B and C combined, it fills in the same time as the two tanks empty, i.e. 14.76 days. The cycle restarts: the rod on float *g* strikes the balanced arm, tilting it towards tank B; valve *p* opens, valve *m* closes and weight *v* is at the top of its travel.

It is assumed that one man, i.e. one man per shift, was responsible for ensuring that the rate of filling and emptying of tanks B and C was constant. If water were erroneously or maliciously extracted from tank C then water was added from tap *a* until the water rose back to its correct level. Similarly, if water were added to tank C, tap *k* was opened until the water fell back to its correct level. It would have been necessary to have had a level gauge in tank B so that discrepancies could be noticed as soon as they occurred, without the observer needing to enter tank C's chamber. This gauge is not shown.

This paper has been an attempt to rescue the Toledo clepsydras from the aura of mystery and wonder-working, an aura which was imparted to them by their Arabic reporter and since then not wholly dispelled by modern commentators. The components and ideas incorporated in this reconstruction were all familiar to Muslim engineers by the eleventh century and there is therefore no reason why they should not have formed part of the Toledo clock. It is not claimed, of course, that the reconstruction outlined here is an exact representation of the original. For example, feedback control could have been achieved as satisfactorily by using an occluded orifice rather than by a float-chamber. The occluded orifice was used by the Bānū Mūsà in a lamp with automatic feed,[15] and in the water-clock described in the *Libros del saber*.[16]

No doubt other alterations could be suggested for some of the components in the water machinery. Nevertheless, it is felt that a system similar to the one shown in Figures 1 and 2 was probably the basis of the Toledo clock. In particular, no alternative to the use of calibrated orifices can be envisaged.

Notes and References

1. *On the Construction of Water-clocks*, trans. and annotated by Donald R. Hill (London, Turner & Devereux, 1976).

2. *The Book of Knowledge of Ingenious Mechanical Devices*, trans. and annotated by Donald R. Hill (Dordrecht, Reidel, 1974).

3. Juan Vernet, *De Abd al-Rahman I a Isabel II* (Barcelona, 1989), 297–300.

4. Donald R. Hill, *Arabic Water-clocks* (Aleppo, Institute for the History of Arabic Science, 1981), 36–46. Balanced pipes are also described by al-Jazarī in his chapter on fountains and musical automata (*op. cit.* 157–77), without claiming that they were his own invention. He mentions that they were described in a treatise by Hibat Allah b. al-Ḥusayn, written in Baghdad in 1123/4. These and al-Ḥusayn used a lead ball instead of the mercury used by al-Murādī, some 70 years before Hibat Allah.

5. M. Rico y Sinobas, *Libros del saber de Astronomía*, 5 vols (Madrid, 1863); see also Silvio A. Bedini, 'The Compartmented Cylindrical Clepsydra', *Technology and Culture*, 3, 1962, 115–41.

6. For example, J. Gimpel, *The medieval machine* (London, Gollancz, 1977), 165–6; David S. Landes, *Revolution in Time*, (Cambridge, Mass., 1983), 77.

7. Hill, *Arabic Water-clocks*, *op. cit.* (4), 47–68.

8. Arīn or *qubbat al-ʿarḍ* is the term used by Arabic geographers to denote the geographical centre of the earth. It is at 90° from each pole and at meridians 0° and 180°, wherever the theoretical zero meridian is located. Its actual location was disputed, but it was said by many authorities to be an island in the Indian Ocean. See *Encyclopaedia of Islam*, 'al-Ḳubba', Vol. V, 1986, 297.

9. *The Book of Ingenious Devices*, trans. and annotated by Donald R. Hill (Dordrecht, Reidel, 1979), 192–209.

10. Al-Jazarī/Hill, *op. cit.* (2), 17–41.

11. Hill, *Arabic Water-clocks*, *op. cit.* (4), 129.

12. The large cisterns, built in Qayrawān, Tunisia, and still existing, use a similar method of decantation; see K.A.C. Creswell, *Early Islamic Architecture* (London, 1958), 291–2.

13. Riḍwān's treatise, describing how he repaired a water-clock built in Damascus by his father, was completed in 1203; see Hill, *Arabic Water-clocks*, *op. cit.* (4), 69–88 (al-Jazarī's comments on the Archimedes clock are on p. 17).

14. Al-Jazarī gives a detailed description of the calibration of orifices. He must have known the approximate diameter required, since he deliberately made the orifice (a piece of drilled onyx) too small, then widened it with copper wire and emery powder until the required rate of flow was accurately obtained. Similar methods must have been used in the Archimedes clock. Vitruvius tells us that the Alexandrian engineer Ctesibius, who flourished at the beginning of the third century BC, made orifices for water-clocks from tubes of gold or by piercing gems — see *De Architectura*, trans. F. Granger (Loeb Classics, Harvard, 1934), 259.

15. Donald R. Hill, *Bānū Mūsà* (Dordrecht, 1979), 236–7.

16. Hill, *Arabic Water-clocks*, *op. cit.* (4), 126–9.

The Other Side of the Coin: Wood Transport Systems in Pre-Industrial Europe

GRAHAM HOLLISTER-SHORT

COAL AND THE ENGLISH

I have prefaced the real title of this paper, Wood Transport Systems in Pre-industrial Europe, with the phrase 'The other side of the coin' for what I hope may be thought sufficient reason.

This has to do with the curious fact that technological writing in the English language is virtually silent on the question – and it is no trivial question – of wood transport. Search as one may in the literature for any reference to this subject, there appears to be nothing except the translation of Johann Beckmann's brief sketch, 'Floating of Wood', relating to some aspects only of the subject, in his *History of Inventions*.[1] Except for the chance occurrence of this piece, I have been able to find nothing at all. It is as if the classic story of English precocity in the mining of coal and England's equally precocious 'take-off' into sustained economic growth on the back of this fuel had quite blinded the English-speaking world to a simple truth: that before the age of coal in England there was an age of wood, and that for continental Europe that truth was (to an overwhelming extent) to remain the truth right through the classical period of the English industrial revolution.

So King Coal might then reign in England but King Wood was still paramount in Europe. Since the levels of industrial production in Europe at that time were maintained at very considerable volumes by reliance on the traditional fuels – wood and charcoal – the quantitative aspects of the situation, that is, questions relating to the acquisition and transport of the vast quantities of wood-fuel required, might seem worth scrutiny. After all, the movement of coal, a much more concentrated form of fuel than wood, from pithead to the great centres of consumption posed major problems that were responsible in large part for stimulating the creation of the canal network. In fact, the movement of wood equally required the invention of methods of distribution and the creation of even more extensive distribution networks across the face of continental Europe. One purpose of this paper, therefore, is to provide a sketch of the nature of those distribution systems, and equally of the very extensive literature that exists in other languages concerned with their history. I hope, in short, to reveal what the other side of the coin was from that smoky image of their great coal-burning past which has, so to speak, concealed from the English the very existence of a European counter-system.

Two recent publications in English, albeit in very different ways, bear on the matters under review here, and a consideration of the nexus of questions they raise may serve as preface to the three-part treatment of the wood transport question which follows. One of these, a recent essay in the *Economic History Review* by Berg and Hudson, 'Rehabilitating the Industrial Revolution', argues that the idea that there was an industrial revolution in England and that it took place in the decades either side of 1800 should be rehabilitated. This had become necessary because the idea that there had really been something properly describable as an industrial revolution in England had, in their words, 'been dethroned almost entirely' in historical writing since the late 1970s.[2] Whether this means that the 'thrice-squeezed orange' still has juice in it or whether one is now down to disputation over pips, I do not pretend to judge.[3] I hope, however, that what is to be presented here on wood transport technology will throw some indirect light on the matter, and that this prefatory section may even shed a beam or two of light upon it directly. The second work referred to above, John Hatcher's *The History of the British Coal Industry before 1700*, is the opening volume in the now completed five-volume series on the history of this basic industry.[4] A fresh look at the period before 1700 is especially welcome since it is now more than 60 years since John Nef published his detailed survey.[5] Although the new study draws upon a much wider range of archival materials than those used by Nef, many of Nef's more important conclusions still stand. By 1700, for example, coal and coal alone was saving England from a pronounced insufficiency of traditional fuels, wood and charcoal, and 'in some instances from a prolonged energy crisis'.[6] By the end of the seventeenth century fossil fuel had in all probability overtaken fuel from self-renewing sources as 'the leading provider of the nation's heat' — the first use of coal as a source of power would begin only in 1712.[7] Meanwhile, the growth of coal production was already, before 1700, achieving rates of increase comparable to those of the industrial revolution period itself, with output running at an annual rate of about 3 million tonnes per year.

What both Nef and now Hatcher have failed to do with this figure, however, is to find some adequate way of transposing it into easily graspable strategic terms. While it is correct to say that coal, unlike wood, 'did not compete for space with grain and animals', this clearly does not take the matter far enough.[8] What did it really mean in terms of actual acreage released for food production, etc. for England to be burning 3 million tonnes of coal a year instead of having to provide the thermal equivalent of so much fuel from forest, coppice and peat digging? Or, more to the point, *how much* arable and pasture, in default of an annual supply of coal of this order, would have had to be sacrificed to maintain the population of England and its economy at the levels then obtaining? The calculation is not in fact a difficult one. It takes a little over 4 m^3 of wood to equal the calorific value of 1 tonne of coal. Four cubic metres of timber equals the annual growth increment, i.e., the *sustainable* take of wood, of 1 hectare (2.47 acres) of good quality forest.[9] It is easy to see, therefore, that to replace 3 million tonnes of coal would have required the setting aside of 3 million hectares, or something approaching 7.5 million acres as forest, to have acquired a resource equivalent to the average annual coal burn of *c.* 1700. A tract of forest of this extent, i.e. 11,718 square miles, would cover 23 per cent of the total land surface of England. A hundred years later coal production had more than tripled. By 1800 it was running at about 10 million tonnes per annum. These figures speak for themselves, so what more need be said? Only, perhaps, to ask how far the English themselves (and outsiders also, for that matter) living through so profound a revolution were becoming conscious of it. As to this, I shall call three witnesses.

The first of these is William Sharpe, whose *Treatise on Coal Mines* appeared in 1769. This small work contains a wealth of indications that the critical difference made by the availability of cheap coal fuel was well understood at that time. Sharpe was writing before the period of canal construction had really gathered pace, so it was easy for him, and others, to observe the contrast between cheap-coal counties and those without any easy access to coal. Indeed, Sharpe cites evidence which suggests that medical statistics were already in existence well before 1769 that pointed to an 800 per cent higher incidence of fever in a coal-less county like Dorset (the result of wearing wet clothes and living in unheated cottages) compared to Durham where firing was cheap.[10] We know from other sources that the only fuel available to the poor in many southern counties (apart from the scraps of wood stolen from fences, gates and hedgerows that Sharpe reports) was cowdung mixed with chopped straw. This mixture was plastered on walls, and, when dry, could be peeled off in thin flakes and used for cooking. Just as the high rates of ill health among men and women pinpointed areas of high-cost fuel, so the same regional pattern was observable in grossly damaged buildings. Speaking of the manufacture of bricks and tiles, Sharpe observed that '[w]here fuel for this purpose [was] dear, the manufacturer [was] always tempted to be frugal with fire', with the result that such inadequately burnt brick '[began] to shiver to pieces'. The truth of this remark was, he added 'evident in every part of England'. Lest this should be taken for some notion of his own, he was careful to add the phrase 'Everyone knows it'.[11] Sharpe's final point in this vein is perhaps his most powerful: 'Why has land trade', he asks, 'of late years made so great a progress northward, and arrived to a pitch of improvement there, unforeseen to the most discerning of our forefathers?' He answers his own question briskly enough. It was '[p]rincipally on account of the plenty of firing'.[12] The concentration of industry on the coalfields was to be a characteristic feature of all early industrializations. Here we see as early as 1769 that English industry was already experiencing this phenomenon, and that, according to Sharpe, at an accelerating rate.

In 1784 Barthélemi Faujas de Saint-Fond toured England. He kept a journal which was, however, to be published only many years later.[13] His opening chapters have largely to do with the personalities he met in London and with giving a sketch of the range of industries which flourished there. Since so many of these industries could only flourish because of the abundance of cheap coal, he was led to comment on the role that coal played in the economic life of England. As he saw it, the incalculable advantages which England drew from coal made it obvious also that coal underpinned the very existence of the nation: 'Le charbon de terre, ce combustible utile, qui tient, pour ainsi dire, à l'existence de l'Angleterre'.[14] A few sentences later, remarking that while it was true that some English travellers in France had tried to awaken the French to the advantages of coal fuel, he felt bound to say that no such openness had characterized the policy of the British government. On the contrary, that government had taken extraordinary care to ensure that France should, as far as possible, for policy reasons, remain in ignorance on a matter which so closely concerned its industry and commerce. Coal was a subject close to Faujas's heart and one to which he promised to return when his journey had brought him to the famous coal-mines of Newcastle. Faujas was as good as his word. It was while he was in Newcastle that he recorded an episode that had taken place some years before in his native Paris, probably in 1782. He tells us how he had gone one day to visit Benjamin Franklin at his house in Passy. The circle around Franklin at that time was indeed a brilliant one, and it was the conversation of that day in Passy that Faujas recalled as he viewed the myriad workshops on the banks of the Tyne. The talk at Passy had turned eventually to the question of how it was that England possessed such power. One of the circle — it was evidently Franklin himself although Faujas declined to name him — had this to say: 'that although no publicist had yet recognised the real reason, there was really no longer any doubt about the cause of English good fortune, it was the coal mines which had worked so many miracles', adding, 'we know that it is almost a treason to the nation to say such a thing in France but then, after all, the French had helped the United States to gain its independence, and moreover I love to see men of every country live happy lives'.[15]

Although we know with the advantage of hindsight that the volume of coal extracted in England year by year through the eighteenth century was doing little more than nibble at the colossal reserves then still largely untouched, the thought of what would happen if the coal-mines were to fail was present in the imagination of one writer at least. This was John Williams. His fears were, of course, unfounded, but the manner of their expression in his *Natural History of the Mineral Kingdom* of 1789 is nevertheless very revealing. Having noted that 'the very existence of the metropolis depends upon the continued abundance of this precious fossil', he saw also that 'when our coal mines are exhausted the prosperity and glory of this . . . island is at an end. Our cities and great towns must then become ruinous heaps for want of fuel, and our . . . manufactures must fail from the same cause.'[16]

HISTORICAL OUTLINE OF WOOD TRANSPORT

The fabric of Western European society suffered severely during the ninth and tenth centuries from the combined effect of sustained Viking, Magyar and Arab incursions. In the face of widespread disorders the population declined severely, with much cultivated land reverting to wilderness as a consequence. With the return of relatively stable conditions in the eleventh century, however, a strong surge in numbers took place, more especially after *c.* 1050. With it came a sustained assault on the forests in order to open up new ploughlands. This assault did not take place undirected. If a landlord, by offering favourable terms to groups of peasants, could persuade them to settle on his domains, this was a certain way to turn forested countryside, hitherto good only for hunting, into farmland producing healthy revenues in the form of rents and dues. Ecclesiastical landlords led the way in this but lay seigneurial landlords soon followed their example. The general term for deforestation at this time was *aedificare*, meaning, literally, 'to build', but more importantly connoting the idea that land was being made profitable. Changes in the law facilitated the process of forest clearance.[17] Legal disputes between those who enjoyed customary rights in the forest and the assertors grew increasingly common through the thirteenth century, a sure sign of the pace at which clearances were taking place. In France, and certainly in many regions of Germany also, the great forests which had once stretched across the countryside from one frontier to another had by about 1300 in many cases been reduced to a mass of fragments.[18]

Parallel with these changes went others of a technological nature, linked in some dialectical fashion with a growing capitalistic or entrepreneurial spirit.[19] It is in the eleventh century that the earliest references to mechanization in some of the basic industries begin to be traceable. As an example of this one might take the iron industry. This industry, a great destroyer of forests, began to be mechanized in the eleventh century. As this took place, so its appetite for charcoal increased. A hand-operated bloomery furnace might produce some two to three tonnes of malleable iron per annum. If, however, the bellows were adapted to be driven by a waterwheel, then production could be increased to some 25 tonnes per year. If the hammer used for deslagging the blooms were also adapted so as to be worked by water power, then output could rise again to about 45 tonnes. Since each tonne of bar iron, that is to say, the finished form in which the iron reached the market, took about three tonnes of charcoal to produce, the enhanced pressure on the forests from this sequence of changes alone can be readily understood.[20] The drive for increased production in the mining of metals, and indeed the rapid extension of mining operations throughout central and eastern Europe through the twelfth and thirteenth centuries, exerted similar pressures on the forests. Vast quantities of timber were needed for mine workings and as fuel for smelting furnaces. The salt industry — for salt was then as vital a commodity as oil is today — was a huge consumer of wood. But whether the need for fuel was generated by domestic needs, for cooking and heating, or for industrial purposes, or as material for ships and houses, it was the forest which was the sole source of provision.

As the number of centers of concentrated fuel or timber consumption grew, whether cities, or salt works, or mines, or smelters, so each would necessarily create problems of supply as stands of timber available in the immediate vicinities of those centres were cut and consumed. In such circumstances, as the radius of provisionment of each inexorably widened, the ordinary means of land transport became increasingly uneconomical and then quite unable to cope physically with the quantities of timber it was necessary to haul.[21] The only response possible (other than passively accepting the situation) was to begin using rivers and streams for the floating of wood.[22] By this means supplies from distant and hitherto unexploited reserves of timber could be brought at small cost to the places of consumption. Such floats might consist either of (i) long, steerable trains or rafts of bound logs or (ii) loose billets, thrown into the stream and thereafter swimming freely with the current. Inland transport of wood by means of ships was also resorted to but seems, judging by documentary evidence, to have been later coming into use than either of the other methods.

The earliest direct references to the rafting of timber on the Elbe for the service of the mines of the Saxon Erzgebirge, although scarcely numerous, indicate that the river carried significant quantities of timber as early as 1325, while indirect references permit the history of this traffic to be traced back to 1292 and, with rather less certainty, to 1177.[23] As for the smaller rivers and streams flowing northward from the Bohemian frontier such as the Zwickau Mulde, the Weisseritz, the Freiberg Mulde and the Floha, it seems likely that all were in use as free-floating streams by the fourteenth century, although it is only through references to it for Zwickau (in 1316 and 1348) that free floating may be traced back so far. However, the Rotengraben (an artificial water course) serving Freiberg was certainly in use before 1400.[24] Thereafter the network of artificial channels (*Graben*) was steadily extended to supplement the ever more ambitious schemes of management to which the natural streams were subjected. What these

Figure 1 The diversion weir on the Mulde at Zwickau with two securing *Steinkasten*. The weir diverted the normal flow of the river down the Muhlgraben. Detail from the Zwickauer Bildrolle 1570 after K Löffler, *Bergbau, Wald, Flösse*, Berlin 1960.

might entail is best illustrated by reference to the Zwickauer Bildrolle of 1570.[25] This document, a colour-washed drawing measuring 60 cm X 618 cm, was produced as a result of a lawsuit brought against the town council of Zwickau by the commune of Scheweditz following the expiration of agreements concerning the boundaries of meadow land. On it are shown the river engineering works along the river Mulde from Zwickau itself to the Kainsdorf weir, that is, the main weir at Zwickau; the rake-works (*Rechen*), that is, wood-catching structures for arresting floating timber on the Mulde and the Muhlgraben; flood diversion channels and bank revetments. Many of the constructional details of these structures shown in the drawing (a unique source of reference for the period) are of the highest interest. One might mention as an example the use of wooden cribs (*Steinkasten*) packed with stones (and presumably also with clay) which served to buttress the banks of the overflow channels and to secure the wooden planking of their flood beds.[26]

But Saxony was but one of a great number of regions whose economic life (in this case largely based on mines and smelters) was sustained by the enormous masses of timber borne to them on rivers, streams and flumes. The great salt works which boiled down the brine of such places as Hall, Hallein and Bad Reichenhall in the Tirol and Bavaria were consumers of wood on a scale which eclipsed even the demands of mining and metal working. At Hallein on the Salzach a rake-work existed already in 1207, a structure which was enlarged and improved in 1494. The fuel needed for the salt pans was derived from forests totalling some 100,000 hectares (250,000 acres) whose annual yield, floated as free-swimming billets along innumerable streams, amounted to something in the region of 140,000 to 160,000 cubic metres: equivalent in thermal value to something like 40,000 tonnes of coal.[27] At Hall on the Inn, likewise, a large rake-work existed in 1307 and was called, significantly enough, like all such structures, the 'work' (*Werch*).[28] By no later than 1400 a similar construction caught

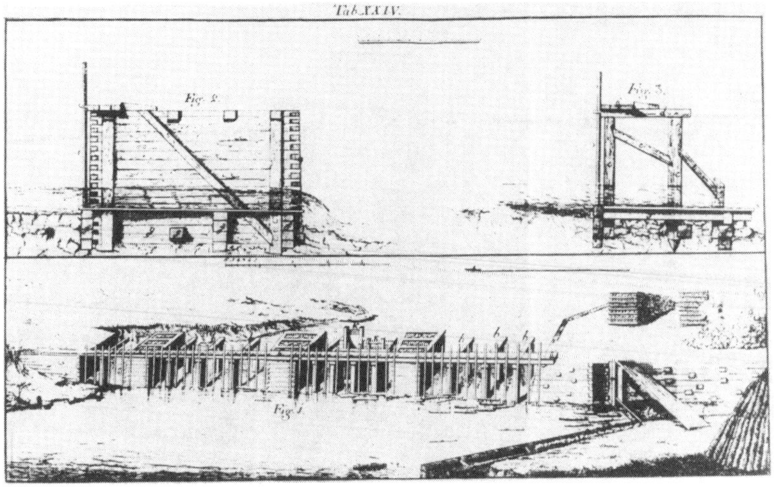

Figure 2 Rechen with Steinkasten.

Source: C.T. Delius, *Anleitung zur der Bergbaukunst . . .*, Vienna 1773, plate XXIV.

the wood carried down by the Saalach to Bad Reichenhall.[29] Here, however, the forest reserves became unequal to the needs of the salt works when a new brine spring was discovered in 1613 whose salt content was nearly as high as that of the old Edelquelle (noble spring).[30] At first an attempt was made to avoid the need for extra fuel (which was in any case unobtainable) by building (in 1615), a structure called variously in German a *Gradierhaus* ('graduation house'), *Strohkunst* ('straw device') or *Leckwerk* ('leak work'), designed to accelerate the evaporation of water, greatly concentrate the brine before boiling and thereby economize on scarce and precious fuel. The graduation technique, in itself evidence of fuel shortage, had by then been in use for something like fifty years in other regions of Germany.[31] At Bad Reichenhall, however, it was, for climatic reasons, not a success, and it was in these circumstances that an idea mooted as early as 1613 was taken up in earnest. This was to pump part of the brine over the mountains to Siegsdorf where adequate reserves of fuel were available. Work was begun on the pipeline in 1617 but it now became part of the plan to continue the line beyond Siegsdorf to Traunstein, lying 30 km to the north-west of Bad Reichenhall. Seven pumping stations (some worked by *Stangenkünste*, i.e. hydraulically powered rod engines, some by horses: *Ross Künste*) forced the brine up to a height of nearly 300 m to Lettenklausen near Inzell whence it ran by gravity to Traunstein. The construction was pushed through with

Figure 3 Brine graduation structure at Bad Söden, Saxony, under restoration in 1986.

such speed that by August 1619 the salt pans at Traunstein were boiling down the first supplies of brine delivered through the pipeline. The daily rate of discharge approached some 200,000 litres.[32]

The great city of Augsburg, with its multiplicity of industries, came to depend for its lifeline at a very early date on the innumerable streams and floatways all of which fed the Lech (called in German the *Lebensader Augsburgs* – Augsburg's Jugular) and the Wertach at whose confluence the city stood. The quickest way to reduce the city to submission in war was in fact to intercept its flow of building timber (*Zimmerholz*) and firewood (*Brennholz*), a tactic employed by Duke Albrecht of Bavaria in his stuggle with the Augsburgers in 1468. But its dependence (via the Lech) on the high forests of Aschau and Reutte in the Tirol had begun long before that as the guilds of float-

Figure 4 End view of the structure shown in Figure 3.

workers in existence in the thirteenth century testify. As for the Wertach, the right of Augsburg to use the stream for floats unhindered and in perpetuity was granted to the city by Abbot Heinrich of Irsee in 1304. On both rivers the floating season began in March and lasted until November or December.[33]

With the passage of time more and more regions came to depend on water transport systems for the conveyance of wood, so that by the sixteenth century what seem to have been originally German techniques (most probably first developed in Styria, the Tirol and the Salzkammergut) were in use virtually throughout Europe. Paris, first beginning to experience severe fuel shortages at the opening of the sixteenth century, started to be supplied with wood by such means from about 1546. Gilles Defroissez in that year was proposing to float wood along the Cure to Cravant, while in 1547 Charles Le Conte opened the Yonne for the floating of billets. Measures like these, designed to exploit the forested mountains of Morvan and Nivernois, later embraced the Saulx and Ornain, tributaries of the Marne, as conveyors of '*bois perdu*' also.[34] At much the same time Pedro Juan Lastanosa, the supposed author of *Los veintiun libros de los ingenios . . .* of *c.* 1570 in a section of Book XV entitled 'De Los Arboles' [On Trees] reveals the already extensive use of such floats in Spain, although the fact that he thought it worth while to suggest ways of improving the courses of natural waterways would seem to indicate an imperfectly understood technique.[35] Edward Browne noted in 1669 that the glass houses of Murano drew their supplies of fuel from the Alps. Wood was floated along a conduit of masonry called the Brentella which ran

Figure 5 Rechen at Banská Bystrica (Neusohl), Slovakia, *c.* 1920.

from Spilimbergo on the Tagliamento south-westwards to the lagoon of Venice.[36] In eastern Europe the important mines of central Slovakia possessed extensively developed water transport systems long before the 1,167-metre-long rake-work across the Gran (Hron) at Neusohl (Banská Bystrica) was constructed in 1547 by an expert from the Tirol.[37]

Europe continued to depend ever more heavily on its forests for fuel until well into the nineteenth century, except for certain restricted regions where coal-mining was well developed.[38] Large-scale engineering works continued to be undertaken to relieve fuel shortages. As an example of the magnitude which such work might attain during this final period of Europe's eotechnic or wood-dependent phase (*c*. 1790–1850) one might take Joseph Rosenauer's scheme to bring the Urwald of southern Bohemia, a terrain of some 300,000 hectares of hitherto unexploited primeval forest, into active exploitation. Vienna, some 250 km to the east and perennially short of fuel, was the market he sought to supply. Rosenauer in 1758 entered the service of Prince Johann Adam Schwarzenberg, who soon recognized his talent as a surveyor and engineer. In 1770 he was appointed forest engineer of the Krumau (Krumlov) forest, a portion of the Schwarzenberg domains. During 1774–5 he made an exact survey of the line along which a flume 49 km long might be constructed beginning at 925 m on the northern slopes of the Dreisesselberg and falling at 790 m into the Zwettlbach, at Rosenhugel. His plan was then to use the already canalized courses of the Zwettlebach (of 7.6 km) and the Grosser Mühl (38.5 km) into which the former flowed, to convey the timber to the rake-work at Partenstein, a little above the confluence of the Mühl and the Danube. Work was begun in 1789 on what came to be called the *alter Kanal*, old flume, a stretch 31.6 km long running from Rosenhugel to Seebach. This was completed and in use by 1791. The upper stretch from Lichtwasser to Seebach was completed long after Rosenauer's death. Rosenauer's original plan for this section of 17 km was to route the flume along the 900-metre contour, but when work began in 1820 it was decided to shorten the line by driving a tunnel through the Flosselberg. This cut over 7 km from the length of the flume by avoiding Rosenauer's projected route following the contour round the Jokesberg. The tunnel, 419 m long, was driven through solid granite in the winters of 1821 and 1822, and by 1824 the *neuen Kanal*, the new flume, 9.7 km in length, was linked to the already existing section at Seebach/Hirschforst. Until 1870 the annual delivery of wood along the route averaged well over 100,000 m^3 (or in coal-equivalent 25,000 tonnes). Even in 1880 an impressively large workforce dressed in military-style uniform (the Schwarzenbergischen Grenadiere as they were called) operated the installations. Two hundred and sixty signallers armed with discs mounted on long poles for announcing the despatch of each new mass of billets, with 650 throwers-in of logs and 320 auxiliaries formed the permanent workforce on the canal.[39]

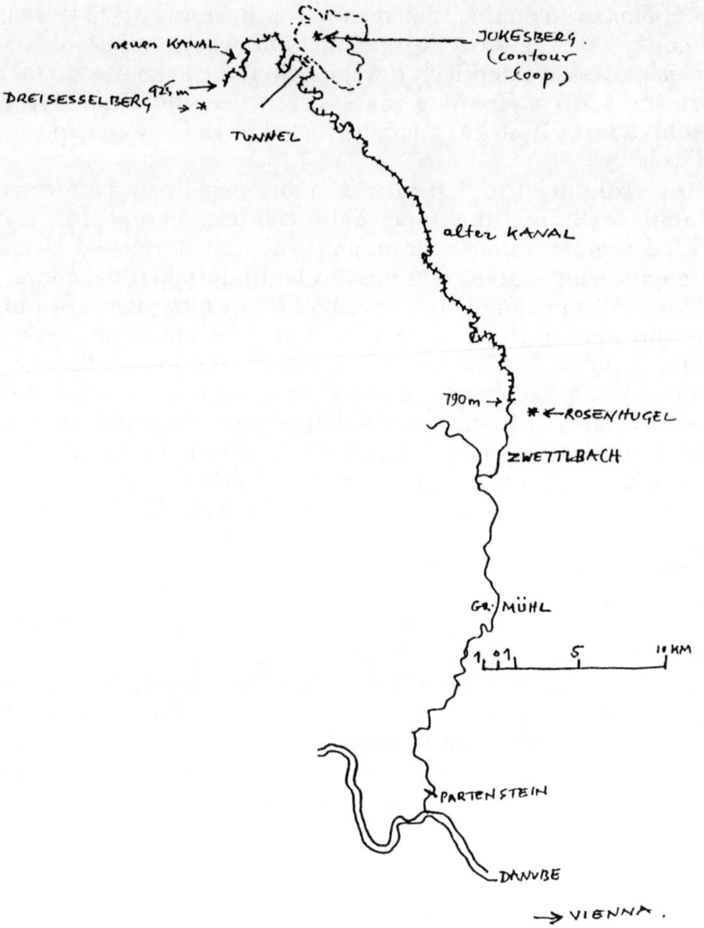

Figure 6 The route of the Rosenauer Schwemmkanal in southern Bohemia.

THE TECHNOLOGY OF WOOD TRANSPORT

As with virtually every aspect of the history of technology, so here also the problem of vocabulary in relation to the techniques of hydraulic engineering as applied to the transport of wood has to be resolved in some degree if the source materials are to be correctly understood. Nomenclature was in fact always highly variable, and according to Neweklowsky such variability is to be found in the records of even the earliest period of wood floating. The word *Trift* for instance (from *treiben*: to drive), which in more modern usage came to signify free-floating wood, was often used interchangeably with *Flotzen* or *Fletzen* (from *fliessen*: to float) whose modern equivalents *Flösse* and *Flösserei* refer more correctly to the rafting of timber. Then again, as far as free floating was concerned, a distinction was sometimes drawn between wood floated on natural watercourses (*Holztrift*) and that conveyed along artificial flumes (*Holzschwemme*: 'swim wood'), although here again both *Trift* and *Schwemm* were used interchangeably at an early period.[40]

As far as water transport of timber is concerned, it is clear that there were three principal modes in use: that conveyed (a) in ships (*Schiffahrt*), (b) in rafts (*Flösserei*), and (c) as free swimming timber (*Trift*). But before the wood actually reached the water it had usually to be transported some distance to it from higher ground. This was done by means of (1) gravity systems, that is, shoots or slideways where the slope was sufficient for the wood to be able to deliver itself; or (2) wooden tracks of gentle gradient (*Winterbahnen, Schmierwege*) on which it was necessary to apply some tractive effort to the timber. Although whole tree trunks (*Bau-Lang oder Zimmerholz*: building or long wood) were usually rafted, this was not invariably the case. Most fuel timber (*Brennholz*) was, however, floated as free-swimming billets for at least part of its journey.

It is obvious that only large rivers could in general be used for *Flösserei*, and the principal works necessary to ensure the passage of trains of timber were specially constructed glideways or flow paths (*Flosgasse*), really inclined planes with guard rails, in the regulating weirs along the rivers which bore such traffic. There were, of course, a variety of techniques for the lashing together of the raft timbers, to say nothing of the special procedures to be followed when wood of different species (and different floating characteristics) were to be floated together.

Figure 7 Rafting timber on the river Vah, Slovakia, *c.* 1920.

As for the smaller streams, although these might occasionally provide a sustained flow sufficient for *Trift* work, it was usually only at the time of spring snow-melt or after heavy rains that most could be exploited. It should be noted here, in anticipation of the next section which examines the situation in more detail, that the heavy cutting of timber in the catchment area of any stream would begin almost at once to affect its regime adversely. Large-scale cutting accelerated the rate of run-off of rain or melting snow, and led consequently to a much attenuated flow after the flush had passed, and to a greatly shortened floating period. For this reason it was usually necessary to construct catchment reservoirs (*Flössteichen*), water from which helped regularize stream flow after the initial flush had passed. Here, as a special case, one should mention that the potential of even the most seemingly insignificant flow of water was unlikely to escape the notice of a float-work expert. Johann Krünitz, discussing this very point, noted (in 1778), '[t]here is in the forest no rivulet (Bach) so small that it cannot be made fit for wood transport'. First one had to build as high a dam as possible to block the flow. The water ponded back eventually formed a pool large enough to hold the billets to be floated. Downslope from this spot similar dams had to be constructed stepwise at intervals of two or three miles. Eventually one reached the rivulet's outfall where it joined a regular stream or river along which wood could be floated without difficulty. These dams, more like flash-locks or sluices (*Klausen*) in reality, were opened in succession as the stock of cut timber arrived so as to maintain the momentum of the float down to the river. The rule of thumb in these matters was that every 2 square feet (1,858 cm^2) of timber needed 2 cubic feet (0.0566 m^3) of water for the job to be done.[41] *Trift* streams (*Triftgewässer*), even if they flowed with sufficient strength, were, unlike raftable rivers, seldom usable in their natural state. A good deal of engineering was usually necessary to bring them into exploitation. Rocks were the worst danger to floating wood and led to a high rate of loss through the splintering of billets. But sandbanks were also very troublesome, as were sinuosities in the watercourse. All had to be remedied. Nor could the stream at low water carry more than 60 to 100 cm of water since otherwise the *Senkholzer* (sunk logs) lying on the bottom could not be easily recovered. Sinkers might amount, according to the buoyancy of the type of wood being floated, from 4 to 12 per cent of the total swim.[42] By degrees as a stream's bed was paved with planks, or its banks revetted with timber, as its course was furnished with cut-offs to eliminate meanders and with floating guide booms to direct billets away from sand banks, it became almost as much a work of art as a completely artificial flume (*Graben*, *Schwemmkanal*) fed by streams and reservoirs. Networks of such flumes draining an entire wood collection region (*Sammelgebiet*) were often switched from watershed to watershed and linked together in arterial ways leading to points of assembly or wood depots (*Holzplätze*). Such were the systems described by Delius in 1773 which served the copper smelters at

Neusohl: *Nebenfluder* (ancillary flumes) ran into the *Hauptfluder* (principal flumes). These discharged their loads of timber into the Gran (Hron), which bore the whole mass down to the Rechen at Neusohl itself.[43] A similar system in use at Hall was described by Feigenpuz in 1707: 'the Inn, into which river this wood is conveyed by the help of mountain streams. These are joined by wonderful works into one torrent which, diverted from one valley to another through a wooden flume, carries [the wood for the salt pans] on its back.'[44]

At the end of its course, whether along stream or flume or river, the floating timber had to be collected. To this end it was necessary to construct a rake-work (*Rechen*). Rake-works were of various forms ranging from a simple row of vertical posts held by top and bottom rails to the massive and elaborate structures necessary on the great rivers. A *Rechen* was usually built at 45° to the direction of flow so as to present a longer collecting length to the stream and prevent any undue accumulation of billets. It was also necessary to select a point where the current slackened so as to reduce the impact of floating wood on the rake. The best rake position of all was behind a curve so that the river itself would tend to drive much of the timber towards the bank and thus lessen the dangerous work of pulling it off the rake itself. Wood delivered in this way was called variously *Scheitholz* (split wood), *Klafterholz* ('six-foot wood'), *Kurzholz* (short wood) and *Brennholz* (fuel wood). *Trift* work was not exclusively concerned with fuel timber, however. The Schwarzenberg Holzschwemmkanal, for instance, was eventually modified to carry *Langholzer* (long logs) of up to 24 m in length.[45]

Figure 8 Wood delivery flume (Fluder, Smyk) near Brezno, Slovakia, *c.* 1920.

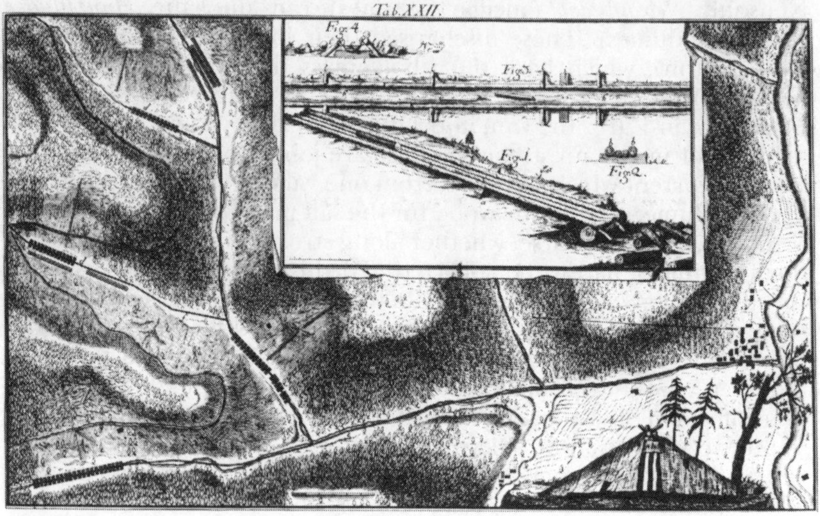

Figure 9 Stocks of cut wood, ancillary flumes and main flume. Note also the wood slide (Riesen), which is labelled 'Fig. 2'.

Source: C.T. Delius, *Anleitung zur der Bergbankunst . . .*, Vienna 1773, plate XXII.

It remains, finally, to consider the various slide systems which served to convey timber cut on the higher slopes down to the valley floors and the water courses which flowed along them. As mentioned earlier, such slides took two principal forms. The first variety were those in which the gradient (usually about 1:8) was sufficient for the billet's own weight to carry it down to the end of the run. They were built of timbers and were given a U-shaped section. At the end of the 'pipe' the timbers were given an upward tilt so as to throw the billets well clear of the point of discharge.[46] These slides went under a variety of names such as *Riese*, *Holzriese*, *Riesel*, *Rutsche* and *Holzglitsche*, and were used mostly in winter when a coating of ice or compacted snow on the slide path timbers added greatly to their efficiency. Some indeed could only be used when they had acquired such a glazing. Others again were operated only when lubricated with water, this to prevent the billets igniting from the friction of the run.

As concerns the second type — those having only slight gradients — it is necessary to distinguish, following the sources, between those used only in winter and those used in summer. The former were called *Winterbahnen* (winter roads). The track consisted of 'sleepers' set at intervals of two feet rather in the manner of a corduroy road with compacted snow filling the spaces. The 'sleepers' were round logs half buried with the bark peeled from the exposed surfaces. Baulks of timber or 'rails' were secured lengthwise to the sleepers which thus had the appearance of a railway. But the rails were simply to prevent the sledges (*Schlitten*) running between them from leaving the track. Normally these sledges carried a cubic *Klafter* (rather over two cubic metres) of split wood (*Scheitholz*). Krünitz (in 1778) describes how if such slides became mushy during the day the work was carried on instead at night by moonlight or by the light of torches. Tracks for summer use (no doubt also winter roads in another season) were called *Schmierwege* ('smearways'). The intervals between the sleepers were filled with tree-loppings while the tops of the sleepers were smeared liberally with fat or grease, replenished after each run. As with regular railways, track junctions were equipped with points.[47]

The decline of all these modes of wood transport set in about the middle of the nineteenth century with the introduction of railways and steam locomotives. The use of coal instead of timber went hand in hand with railway construction. Vienna, for instance, which had consumed 720,000 m^3 of wood fuel in 1815, was burning only 187,000 m^3 by the late 1880s.[48] Rather more slowly the rafting of timber also lost ground to the railways.[49] But there is a curious final twist in the story. As coal replaced wood fuel and as all over Europe demand for fuel rose far above the levels the forests could have sustained, the demand for timber remained colossal. In the form of pit-props the abundance of wood was a primordial necessity for the exploitation of coal. Every tonne of coal extracted 'cost' 0.25 m^3 of timber.[50]

THE ECOLOGICAL CONSEQUENCES OF DEFORESTATION

The treeless, devastated landscapes depicted by the illustrators of Agricola's *De Re Metallica* were no artistic licence, and it is clear that the continually intensifying exploitation of the European forests could not but affect the hydrological balance of whole regions. The devastating effects of the technique, described in the previous section, for sluicing timber down mountainsides can well be imagined but would be as nothing to the results following the scouring away of topsoil left exposed by deforestation. Despite the great interest the subject possesses, it is not one which any historian has, I believe, yet attempted to trace. It is easy to see that such a study presents enormous difficulties since the evidence would, no doubt, be of a highly scattered and incoherent nature. Materials exist, however, which make it evident that by about 1700 at the latest records of river behaviour were being kept. The recorders were concerned in particular with such matters as flood heights and with the ever-decreasing volumes of water flowing along rivers in summer.

The secular effects that inspired such record keeping were not insignificant. It was observed that the frequency of devastatingly high floods was greater than it had been in the past, and yet, paradoxically, although this seemed (to the recorders) to constitute evidence that more rain fell than formerly, the fact remained that in summer the rivers sank to lower and lower levels, hindering traffic and sometimes preventing it altogether. These were serious consequences, for it was the major rivers which were under observation and seen to present increasingly disturbed regimes. But what was the cause? Although there was much evidence to suggest that deforestation was responsible, it was not until the second half of the nineteenth century that any serious experimental study was carried out which attempted to quantify the effects on a watershed of a large-scale clearance of timber. In 1860, however, F. Jeandel, J.B. Cantegril and L. Bellaud published in the *Comptes-Rendus* of the Académie Royale des Sciences their *Études expérimentales sur les inondations*.[51] Despite the severe criticisms levelled against its methodological shortcomings, J. Vaillant's report to the Académie on the *Études expérimentales* was far from unfavourable. It was clear to Vaillant that a most important area of scientific research had been opened up which, taken further, would lead by degrees to a clear understanding of how the important task of regulating water flow was to be handled. What the authors of the *Études* had been concerned to demonstrate, however, was that forests exerted a considerable damping effect on stream flow: that the flood crests of the forest streams in their experimental control area had been lower than those of streams in the cleared zone which they had studied. Their response to Vaillant's criticism, which indicated that for him this finding had still to be regarded as an open question, contains a number of interesting remarks. (The *Études*, Vaillant's report and an appendix by the authors replying to his criticisms were published together in 1862.) They noted first that their study had shown beyond question that the duration of run-off of rain from their forested area was double that of the cleared zone, while in times of flood the streams of the cleared zone had more than doubled in volume compared with the forest streams. They concluded that forests had the effect of reducing by half the danger of violent flooding action from streams. In support of their experimental findings they appealed to observable facts. They cited the well-known case of the commune of Labrugnière (Tarn). Before the Revolution it had possessed some 1,834 hectares of forest. Following the revolutionary legislation which removed the penalties attaching to unauthorized cutting of timber, the forest had been so devastated by 1840 that it was capable of producing only trifling quantities of brushwood. As the forest had disappeared, violent floods had become more and more frequent, forcing the mills to stop working, while in summer the briefest dry spell reduced stream flow to such an extent that the mills were once more repeatedly forced to stop operations. Remedial action, that is reafforestation, had begun in 1840 and although an area of 150 hectares was still waste, experience had shown that *pari passu* with the replanting of forest it had become possible to work the mills more regularly and that sudden violent floods had gradually ceased to occur. It had been noted that water levels in the Caunan began to rise only some six to eight hours after the beginning of rain and that their flood crests followed a regular pattern in coming to and declining from their peaks. They concluded:

> 'Cette exemple est remarquable en ce sens que, toutes les autres circonstances étant restées les mêmes, on ne peut attribuer qu'au reboisement des changements survenus dans la régime du cours d'eau, changements qui peuvent se résumer en deux mots:
>
> (1) attenuation de la crue au moment des pluies;
> (2) augmentation du débit en temps ordinaire.
>
> On comprendra, d'après l'exemple qui précède, le rôle capital que sont appelées à jouer les montagnes couvertes de forêts. En retardant l'écoulement d'une partie les eaux au moment des pluies, elles diminuent les chances d'inondation. En augmentant le débit des cours d'eau en temps ordinaires, elle peuvait rendre de grand service au point de vue de l'agriculture et de l'industrie. A ces avantages, déjà si précieux, il faut ajouter encore celui d'un accroissement notable de production de combustible.'

But in any case, the authors concluded, travellers of such authority as Humboldt, Boussingault and Becquerel had recognized that wherever clearances had taken place on a large scale, whether in Europe, Asia or America, the normal volume of flow in watercourses had diminished, that it built up again with reafforestation, and lastly that there had been no such variations before the forest cover had been disturbed.[53]

The difficulty of setting up experiments of the kind that Jeandel, Cantegril and Bellaud had carried out are obvious enough. How easily can one find two catchment areas as nearly comparable as possible, both of which should initially be forested but one of which, after the first series of recordings, should be cleared of timber so as to permit a second series to be carried out?

After the *Études expérimentales* of 1860 it appears that only two further experimental studies have been published, that of A. Engler in 1919 on the Emmenthal in Switzerland, and that of C.G. Bates and A.J. Henry in 1928 on Wagon Wheel Gap, Colorado. The final report of Bates and Henry's study, 'The Forest and Stream Flow Experiment at Wagon Wheel Gap', summarized the results of an exhaustive series of recordings extending from 1911 to 1926.[54] The two basins they had studied presented virtually identical features. Records were kept of basin A and basin B from 1911 to 1918, when both were forested and then again from 1919 to 1926, after B had been almost totally cleared. The watersheds of both were of 200 acres. The mean air temperature of B rose from 34° to 35.3°F, maximum readings were up to 2.1° warmer and minimum 0.7° colder. Soil temperature rose by up to 3.4° at a depth of 12 inches. Mean wind speed increased by 2.6 m/s, and humidity dropped. What is of the greatest interest, however, was that a very large alteration took place in the speed of run-off. Despite the fact that the soil was coarse and had a high natural capacity for storage, flood discharge at snow-melt increased by over 25 per cent, total discharge went up by 15 per cent and the ratio of peak flow to low increased from 12:1 to 17:1. Flood crests in basin B, formerly only 6 per cent greater than in A, rose on average to 64 per cent after clearance. The rapid flush of water in April and May meant that water was lost which later in the year might have served for irrigation. The '*deux mots*' of the *Études expérimentales* apply almost perfectly. As regards scouring of the terrain, the nature of the soil in basin B precluded any possibility of destructive erosion, yet even so the amount of silt carried by its streams rose by eight and a half times, and, if periods of flood peak are taken alone, by twelve times. Dams were therefore not an attractive proposition. The loss of soil per acre per annum rose from 2.8 lb before clearance to 16.7 lb afterwards, a ratio of 1:6. The authors remark that these findings were untypical. Geological surveys of other cleared areas of the United States had shown that soil loss ratios of 1:50 were not uncommon, and that readings as high as 1:150 had been known to occur, with correspondingly augmented levels of erosion.

What such studies indicate is that the eotechnic phase of European development must, without question, have been increasingly destructive of the environment, a fact which has indeed been scarcely recognized but which can have been no less damaging in its effects than those accompanying the use of coal. The scouring away of topsoil, the conversion of streams into torrents, the rapid silting up of reservoirs, to say nothing of the reduced possibilities for transport and industry and the general devastation caused by floods, were the penalties attaching to an exclusive dependence on wood fuel, where production was pushed beyond the limits prescribed by the teachings of *Forstwissenschaft* (forestry science) and by reasonable care for the fuel supplies of posterity. The patient recension of local histories would surely reveal in intricate detail the pattern of these events.[55]

Notes and References

1. J. Beckmann, *History of Inventions*, trans W. Johnston (London, 1877), Vol. 1, 454–63. The work was first published in Leipzig 1780–1805 in 5 vols under the title *Beyträge zur Geschichte der Erfindungen*. M.J.T. Lewis touches on the matter peripherally in his *Early Wooden Railways* (London, 1970), 69–70, in searching for precursors of the guide-wheel railway. He considers briefly the possible role of slide-ways (*Riesen*) and (with more plausibility, in my opinion) what I shall describe in the second section as winter-roads and smear-ways.

2. M. Berg and P. Hudson, 'Rehabilitating the Industrial Revolution', *Economic History Review*, 1992, Vol. XLV, No. 1: 25.

3. The phrase 'thrice-squeezed orange' was first quoted by J.H. Clapham. See *The Economic Journal*, 1910, Vol. XX: 195. 'Even if, as a great economist once said to me, the history of "the" industrial revolution is a "thrice squeezed orange", there remains an astonishing amount of juice in it.'

4. J. Hatcher, *The History of the British Coal Industry*. Vol. 1, *Before 1700: Towards the Age of Coal* (Oxford, 1993).

5. J.U. Nef, *The Rise of the British Coal Industry*, 2 vols (London, 1932).

6. J. Hatcher, *op. cit.* (4), 7.

7. *Ibid.*, 8. The first use of coal for power in 1712 refers to the fact that steam for Newcomen's first atmospheric engine, built in that year, was raised in a coal-fired boiler.

8. *Ibid.*, 6.

9. G. Huffel, *Économie forestière* (Paris, 1910), Vol. 1, Part 1, 12.

10. W. Sharpe, *Treatise on Coal Mines*, (London, 1769), 10–11. Apropos the stealing of wood, one might mention that in the law codes of several German states in the seventeenth century the penalty for stealing wood from swim canals was death.

11. *Ibid.*, 16.

12. *Ibid.*, 19.

13. B. Faujas de Saint-Fond, *Voyage en Angleterre . . .* (Paris, 1797), 2 vols. An English translation appeared in 1799 and was reissued in revised form in 1907.

14. *Ibid.*, Vol. 1, 130.

15. *Ibid.*, 178. Franklin was then in Paris serving as plenipotentiary of the United States in its negotiations for a peace treaty with Great Britain.

16. J. Williams, *Natural History of the Mineral Kingdom* (Edinburgh, 1789) Vol. 1, 160, 172.

17. On this subject see E. Bonvalot's splendid study, *Le Tiers État d'après la charte de Beaumont* (Paris, 1884), Chs I, II and III. The text of the charter is printed on 98ff. This charter of 1182, granted to the settlement of Beaumont by Archbishop William of Reims, is a landmark in the process by which serfs were made free men. It granted them virtually free use (clause 54) of the seigneurial forests. See also G. Huffel, *op. cit.* (9), Vol. 2, Part 1, 137.

18. G. Huffel, *op. cit.* (9), 141.

19. As to this enormously important but largely neglected question, see J. Baechler, *The Origins of Capitalism* (Oxford, Blackwell, 1975), and L. White, *Machina ex Deo: Essays in the Dynamism of Western Culture* (Cambridge, Mass., 1968). Here one may say that in the work of St Thomas Aquinas (1225–74) the state is largely governed by the economic motive. The basic features of bourgeois values have become clearer still in Duns Scotus (1266–1308), who establishes the principle of the freedom of contract, and in Jean Buridan (1300–58), in whose work the doctrine appears that a morally good man who cared for the common weal ought not to be hindered from growing rich since he brought benefits to the community.

20. H.J. Schubert, *History of the British Iron and Steel Industry* (London, 1957). It took 4 tonnes of timber to produce 1 tonne of charcoal.

21. In England the coal haulage was not economically practicable beyond about ten miles. J. Hatcher, *op. cit.* (4), 14. As late as the eighteenth century land haulage grew ruinously expensive even for manufactured products beyond a very few miles: cf. P. Mathias, *The Brewing Industry in England 1700–1830* (Cambridge, 1959), xxxii: 'With the high overland transport charges of the pre-railway age ordinary strong beer in the 18th century had an economic marketing area, by land carriage, of only 4 to 6 miles.' And this was a high-value product.

22. It is precisely the nature of the choices the West made and the sustained character of its determination to overcome obstacles to action that make it meaningful to speak of the dynamism of Western culture, whatever may have been the system of values, or the 'calculus' which inspired it.

23. H. Wilsdorf, W. Herrmann and K. Löffler, *Bergbau: Wald: Flösse* (Berlin, 1960), 112.

24. *Ibid.*, 107. The Mulden Flösse (*Graben*), also serving Freiberg, was constructed in 1438. Altenberg was served before 1464 by the Aschergraben. Many more works of this sort were undertaken in Saxony in the sixteenth and seventeenth centuries. As the seventeenth century wore on, however, the demands of the mines for water for hydraulic engines became so great as to cut short further flume development for fuel transport.

25. The drawing is preserved in the city archives, Zwickau. Portions of it are reproduced in colour by K. Löffler, 'Die Geschichte der Frieberger Muldenflösse . . .', in H. Wilsdorf *et al.*, *op. cit.* (23). See especially 81, fig. 29; and 83, fig. 31.

26. See also C.T. Delius, *Anleitung zu der Bergbaukunst nach ihrer Theorie und Ausubung nebst einer Abhandlung von den Grundsatzen der Berg-Kammeralwisenschaft* . . . (Vienna, 1773), 514 and plate XXIV (fig. 1). Where a *Rechen* was of some length *Steinkasten* were to be set at intervals along it. A. Solzhenitsyn, *Gulag Archipelago* (London, 1976), Vol. 2, 86, recounts that wooden cribs of this sort, up to fifty feet high, were built to support the lock chambers on the Belomor (White Sea) Canal in 1931–3.

27. E. Neweklowsky, *Die Schiffahrt und Flösserei im Raume der Oberen Donau* (Linz, 1964), 538–9. The *Rechen* at Hallein was one of the largest in Europe.

28. *Ibid.*, 538.

29. *Ibid.*, 541.

30. M. Flurl, 'Ältere Geschichte der Saline Reichenhall vorzüglich in technischer Hinsicht bis zur Erbauung der Hülfs-Saline Traunstein', *Denkschrift der königlichen Akademie der Wissenschaften* (Munich, 1811), Vol. 2, 183. The new spring called the Plattenfluss (silver fountain) delivered something over 400 gallons (1,800 litres) a minute.

31. The idea was to accelerate evaporation of weak brine by passing it repeatedly through layers of straw (*Stroh*) or, later, brushwood. The earliest reference to the technique occurs in 1559 when Caspar Seeler, mint master at Augsburg, made an offer to Duke Albert V of Bavaria to set up at his own expense a work which would greatly reduce the amount of fuel needed to boil down brine. Such an offer seems to me a sure sign that the device he had in mind was then in its experimental stage. According to Virgil Hoffer, the controller of the Bad Reichenhall salt works at that time, Seeler's plan was to build a *Luftgradirung* (air graduation house) for the preliminary concentration of weak brine. At Bad Soden, near Frankfurt am Main, graduation houses raised a weak brine of only 1.5 per cent (scarcely brackish to the taste according to Dr Gilbert Burnet who visited there in 1685) to 35 per cent before boiling. A *Leckwerk* was in use at Nauheim, Saxony, in 1579. Graduation houses were commonly of enormous size. That built at Rosières (Lorraine) in 1739, for instance, was over 3,000 feet long, 45 feet high and 26 feet wide.

32. The surveying and levelling of the line were carried out by Tobias Volcksmer, who presented his draft to the ducal commission in June 1616. The building of the installation in 1617–18 was the work of Hans Reiffenstuhl. The survey plan of the pipeline (a copy of the original made in 1755) shows numerous massive trestle structures, water towers and cuttings. The survey and plan are both preserved in the archives of the Deutsches Museum, Munich.

33. E. Neweklowsky, *op. cit.* (27), 238–44.

34. M. Devèze, *La vie de la forêt française au XVIe Siècle* (Paris, 1961), Vol. 2, 31–8. C. Caron, *Traité des bois servans a tous usages* (Paris, 1676), Vol. 1, 107, 'on jette le bois bûche a bûche dedans [in some "ruisseau assez raissonable"] et c'est ce que l'on appelle à bois perdu, qui suit le fil de l'eau jusques a ce qu'il soit arreté au lieu destiné pour estre mis en train . . .' Sweeping powers were given in the articles of the French *Ordonnance* of December 1672 to engineers of float works to use any water for such work, with liberty to overturn all other forms of property right which might hinder such work.

35. P.J. Lastanosa, *Los veintiun libros de los ingenios* . . . (Madrid, *c.* 1570), f249r. 'Y he pensado una cosa que me pareze que se podria servir del tal rio [strewn with many rocks — "muchas penas"]; yr entablando un camino el quel friese tan ancho que un madero pudiese caminar libramente': 'And I have thought of a way which appears to me would be of service in such a river [strewn with many rocks]; that is to construct a flume [literally, path or road] of sufficient width for a log to be able to travel freely', i.e. wide enough for logs to spin without catching the sides. Lastanosa goes on to caution against building a flume with tight curves. I have to thank Dr Alex Keller for the transcription of this passage. The translation is my own. Although the authorship of this MS has been and is hotly disputed, it seems to me that the case for Lastanosa as author has been strongly made by N.G. Tapia, 'Y sin embargo es Lastanosa', *Tecnica Industrial*, 1991, No. 203, 54–61.

36. E. Browne, *A brief account of some travels in Hungary, Austria, Servia . . . Carniola and Friuli* (London, 1673), 137, 'I passed a neat river or notable cut called *La Brentella*, sixteen miles long . . . for the better bringing down of wood from the mountains to be used in the making of glasses at Muran; it is all paved with a good stone bottom and sides . . .' I have not been able to discover when this was built but I suspect that the notable increase in woollen textile manufacture in Venice after about 1550 may well be connected with improvements in fuel supply. Browne also noted similar works serving the mercury mines of Idria in Friuli.

37. *Ibid.*, 105, 'there is a bridge of wood to pass the river at this town, and an handsome building of piles . . . to stop the wood thrown into this river ten miles higher where the country is full of wood; and by this artifice without labour or charge it is conveyed to Neusohl'. Browne visited Neusohl in March 1670. See also C.T. Delius, *op. cit.*. (26), 514, and for Slovakia especially P. Deffontaines, 'La vie forestière en Slovaquie', *Travaux Publies par l'Institut d'Études Slaves* (Paris, 1932), Vol. XIII, Chapter 5, 44–55, especially 'Les Smyky', 49–50. '*Smyky*' is the Slovak word for '*Fluder*'. There were some 170 kilometres of them around Banská Bystrica in 1926.

38. Principally the region extending from Aniche to Aachen, but many other small fields were worked in France and Germany. By the 1740s and 1750s there was, in both countries, a growing consciousness of the need to expand the production of coal. In France this is marked by Louis XV's Arrêt of 1744 resuming regalian rights over all coalfields.

39. I have drawn the details for this sketch from L. Hauska, 'Joseph Rosenauer und sein Werk', *Zentralblatt für das Gesämte Forstwesen* (Vienna, 1940), Vol. 66, 87–99; E. Neweklowsky, *op. cit.* (27), 508–83; and J. Blau, 'Böhmerwalder Haus-industrie und Volkskunst', *Beiträge zur deutsch-böhmischen Volkskunde* (Prague, 1917–18), Vol. XIV, 1917–18. On p. 71 Blau has a charming photograph of a lady signaller (with pole) standing by her 'signal box'. The Bavarian government ordnance survey maps are undispens-

able if one wishes to trace the details of Rosenauer's survey. The dimensions of the Schwemmkanal, trapezoidal in form, were from 3.8 to 5.7 m in width at the top, 1.9 m at the bottom and 96 cm deep. There is also, of course, much modern literature on the subject in Czech, which however is so extensive that I shall not cite it here.

40. E. Neweklowsky, *op. cit.* (27), 527–8. Saxony had a peculiar nomenclature of its own. *Flösse*, which elsewhere signified rafting, was here used for the free floating of wood and was even extended to include wood delivered by lubricated slideways. The word for rafting having been pre-empted in this way, it became necessary to use the awkward term *Flöss-Schiffahrt* (raft-ship transport) in its place. See also H. Wilsdorf *et al.*, *op. cit.* (23), 56.

41. J.G. Krünitz, *Oeconomische Encyclopädie oder Allgemeines System der Staats-Stadt-Haus-und-Landwirtschaft* (Berlin, 1778), Vol. XIV, 291. 'Es ist in den Waldungen ein Bach so klein der nicht zur Holzflösse gebraucht werden könnte.' See also C.T. Delius, *op. cit.* (26), 511 and plate XXIV, whose entire section on forest economy (pp. 508ff.) is of the highest interest.

42. H. Wilsdorf *et al.*, *op. cit.* (23), 57. The floating characteristics of the different species of wood had also to be considered. Beech would not swim more than 12 miles, and birch, which swam well at first, became saturated and sank after 24 hours. Most varieties could not be floated in their freshly cut state: hence the need to stockpile billets well before the floating season. The French term for *Senkholzer* was *bois-canards*.

43. C.T. Delius, *op. cit.* (26), 510–11.

44. J.E.O. Feigenpuz, *Iter per salinas Tyrolenses* (Innsbruck, 1707), 303. 'Oenus, in qē ope rivulorem miro artificio e montibus in unum torrentem collectorum, et per ligneos alveos subjectos ex una valle in aliam derivatorum haec ligna deportantur suo tergo subvectat'.

45. L. Hauska, *op. cit.* (39), 98.

46. C.T. Delius, *op. cit.* (26), 509 and plate XXII, figs 2 and 3. The upturned end was called a *Sprung*. Apropos of billets landing well clear of the end of the 'pipe', I cannot forebear to mention the famous iron pipe at Urach in Wurtemberg noted by Johann Keysler about 1729. It was over 900 feet long, and the billets, almost like bullets, shot out of its mouth at such speed that they flew ('in freyer Luft hinausfahrt') over 200 paces before landing. Ordinary slides were sometimes up to a mile in length.

47. J.G. Krünitz, *op. cit.* (41), 296ff. and figs 772 (*Schlitt*) and 773 (*Winterbahn: Schmierweg*).

48. E. Neweklowsky, *op. cit.* (27), 531–2. 'die mineralische Kohle das Holz als Brennmittel allmahlich verdrangten, was mit der Erbauung der Bahnen hand in hand ging'.

49. *Ibid.* 236, 'Der Rückgang der Flösserei begann mit der Mechanisierung des Verkehrs durch die Einführung der Dampfmaschine'.

50. G. Huffel, *op. cit.* (9), Vol. 1, Part 1, 19, establishes this equation, deriving his information from French, German and US coal-mining studies and statistics.

51. F. Jeandel, J.B. Cantegril and L. Bellaud, *Études expérimentales sur les inondations* (Paris and Nancy, 1862), 139, note 1.

52. *Ibid.*, 140, note 2.

53. *Ibid.*, 141–3. The quoted passage may be translated as follows: 'This example is remarkable in the sense that, in the absence of all other changes, one can attribute only to reafforestation the changes that followed in the regime of the stream, changes which may be summed up in two words: (1)

lengthening of flood time during rainfall; (2) increased flow in dry weather. One will understand from the example given above the very important part played by forested, mountainous catchment areas in all this. By slowing down the proportion of water flowing off at the moment of rainfall, they reduce the likelihood of flooding. In increasing stream flow at other times they are able to provide an important service to agriculture and industry. Apart from these already valuable benefits, one must not forget yet another: a notable increase in the production of wood for fuel.'

54. C.G. Bates and A.J. Henry 'The Forest and Stream Flow Experiment at Wagon Wheel Gap', *Monthly Weather Review*, No.30, Washington, 1928, 1–79. I should mention here J.H. Eves's 'Shrunken to a comparative rivulet: reforestation, stream flow and rural milling in 19th century Maine', *Technology and Culture*, 1992, Vol. 33, No. 1, 38–65. This might seem to throw some doubt (for Maine anyway) on the validity of Bates and Henry's work, although Eves seems not to know of it.

55. In 1767, for instance, the Augsburg senate records state that the Wertach had by then become scarcely floatable. The collection of codes of forest laws such as those of N. Meuer, *Jagd und Forstrecht . . .* (Frankfurt, 1576), and A. Fritsch, *Corpus juris venatorio-forestalis, Romano-Germanici, tripartitum* (Leipzig, 1702), constitute an archive in themselves. It is clear also from A. Surell, *Étude sur les torrents des Hautes Alpes* (Paris, 1841), that French forest laws would prove an equally valuable fund of source material.

The Blast Furnace and the Mass Production of Armour Plate

ALAN WILLIAMS

INTRODUCTION

No figure is more evocative of the Middle Ages in Europe than the 'knight in shining armour', but the familiarity of this image tends to blind us to its singularity. No other culture produced suits of homogeneous armour made of plates—India, China, and Japan used mail or lamellar armour made of interlinked rings or small plates with overall flexibility. Although such armour offers a good defence against missile weapons, it is inflexible and cannot easily be passed from one wearer to another; indeed the notion of armour as being made for one customer individually is an example of an affluence in manufacture that was itself unique at that time.

Suits of armour were made and used for some two and a half centuries, and then passed out of use for a variety of reasons which are still not wholly understandable.

Those scenes from Olivier's film, *Henry V*, which show the French knights being hoisted by cranes into their saddles, and then shot down by the English archers on the field of Agincourt, have so permeated the collective consciousness that for many people they epitomize the Middle Ages and its military technology.

On reflection, of course, it is clear that armour must have been perceived as offering enough protection against arrows to make its purchase worth while; indeed, Agincourt was (with the possible exceptions of Verneuil (1424) and Flodden (1513)) the last major victory to be won by English archery, and the highpoint of armour was to lie well into the future in the middle of the sixteenth century.

What made these knights and their armour obsolete? A popular answer might be gunpowder, but again the causes of change are complex; while there were hundreds of guns on the battlefield of Agincourt, armoured men were still, some two hundred years later, engaged in the Thirty Years' War (1618–48). Throughout this 250-year period, guns and armour developed in mutual competition. Both armour-piercing guns and bulletproof armour were made until eventually, in the mid-seventeenth century, it was armour that fell into disuse.

MAIL

The precursor of plate armour was mail. The oldest piece of interlinked mail yet found was excavated from a third century BC Celtic grave in Romania.[1] This was probably developed from protective garments made up of rings threaded onto cords, like netting; a fragment of such a garment was found in a much earlier Celtic grave in Bohemia, perhaps of the eighth century BC.[2] It is described as a Gallic invention by Roman writers such as Varro[3] in the first century BC, and Roman soldiers of that period are depicted on, for example, the altar of Domitius Ahenobarbus[4] as wearing mail.

The Roman legions were composed of infantry generally armoured with a shirt of iron mail, an iron helmet, and a large shield, and armed with a pair of javelins and a short sword made of steel but not necessarily quenched. Their officers frequently wore armour of a traditional Greek pattern, and a number of bronze helmets have survived. The legions were organized into a more flexible disposition than the Greek phalanx, and, accompanied by auxiliary (unarmoured) cavalry who played a subordinate role, they found no enemy to defeat them in the Mediterranean world.

The political abilities of the Romans enabled them to consolidate their conquests into an empire that survived for 400 years, and longer still in the eastern (Byzantine) part.

The most famous series of pictures of Roman legionaries, are those carved on Trajan's Column, of about AD 113, which shows them protected by the '*lorica segmentata*', or a type of laminated plate armour, which had gradually replaced mail during the first century AD, and seems to have remained in use until the third century. Other forms of armour remained in use however; the column of, for example, Marcus Aurelius shows legionaries in both mail and plate, and scale and lamellar armour is also depicted. Lamellar armour can be traced back to the Assyrian period and consists of small plates laced to one another (but not to the underlying cloth garment) for the greatest flexibility. Excavations at Dura Europos in Syria uncovered a bronze scale horse armour from the middle of the third century AD. Some fragments of a scale armour (*lorica squamata*) excavated from the Roman fort at Corbridge were found to be a brass of composition 65% Cu, 13%Zn, 2%Sn.[5] Byzantine emperors are shown in cuirasses of lamellar, or sometimes scale type, but never in the *lorica segmentata*, so the period of its use seems not to have been very prolonged.

Throughout the Migration Period and into the early Middle Ages, the mail-shirt (byrnie, hauberk) was the principal body defence for those Germanic warriors fortunate enough to be able to afford it. Its adaptability, however, meant that it was frequently repaired and reused, so that very little has survived from this period, even allowing for the mischances of war and corrosion.

Rust stains were all that was left of a mail-shirt buried with an Anglian king at Sutton Hoo in the sixth century. An Anglo-Saxon helmet of the eighth century was excavated from Coppergate, York, and has been studied recently.[6] It was made of several plates of low-carbon iron riveted together to form a skull, with a nasal guard, from which a curtain or aventail of mail was suspended consisting of alternate rows of riveted and butt-welded links. These also consisted of iron of variable (generally 0.2%C or less) carbon content. No attempt had been made to harden this by heat treatment.

The earliest mail shirt which is relatively intact is that said to have belonged to St Vaclav (or Wenceslas, d. 935) in Prague Cathedral.[7]

The Vikings also seem to have employed mail. A series of boat-burials at Vendel in Sweden, Valsgarde, and elsewhere from the seventh to the ninth century have yielded samples of mail with reinforcing pieces of plate.[8] This innovation might be ascribed to an Eastern influence, or, conversely, Scandinavians might have influenced the Near East. Certainly, Vikings employed the rivers of Russia to trade to the Black and Caspian Seas as well as setting up the kingdom of Kiev, from which nucleus medieval Russia grew. Russian armour of the Middle Ages was a mixture of mail, scale and lamellar[9] not dissimilar to that depicted in Byzantium.

In the West the Vikings settled in northern France and adopted a Latin language and the Roman Church. They also adopted the Frankish style of fighting on horseback — although it had not enabled the Frankish kings to keep them out — and when their Duke invaded England in 1066, employed it to defeat the Anglo-Saxon foot-soldiers.

The Bayeux tapestry shows Normans and English wearing what are essentially similar mail shirts extended to cover the elbow and the knee, while MS illustrations of the twelfth century show the sleeves extended to the wrist, matching leggings to protect the legs, and hoods of mail like balaclavas over the head and neck.

While the body defence made entirely of mail reached its peak of perfection in the thirteenth century, its limitations were becoming apparent when it faced missile weapons, and the additional protection of plates was found desirable. At the battle of Tagliacozzo (1268) the German knights wore defences beneath their mail which rendered them invulnerable to the swords of the French until the Germans reaised their arms to strike. However, the extra weight of these plates exhausted the Germans and made them easy prey for their enemies.[10]

The knees and shins, the parts of a horseman most vulnerable to sword and axe cuts from infantrymen, were protected by metal reinforcing plates from the middle of the thirteenth century, and arm defences were added later in the century. The 'coat of plates' became common probably after the second quarter of the fourteenth century; this was a defence for the upper body made of overlapping plates of iron, or sometimes plates of *cuir-bouilli* (boiled leather)[11] or horn, attached to a fabric jerkin and worn over the mail shirt. Fragments from some 24 coats of plates were excavated from the mass grave at Vishy after the battle of 1361, as well as quantities of mail.[12]

A simpler, cheaper form of the coat of plates, with smaller scales, the 'brigandine' remained a cheap and popular form of defence until the late sixteenth century, but for those who could afford the best, the coat of plates was to evolve into a more complete defence.

The plates became larger in size but fewer in number, until they could be regarded as a rudimentary breastplate with some smaller plates attached. Such breastplates were still worn over a mail shirt in the closing years of the fourteenth century – and one survives in the castle of Churburg in the Tirol – but the logical conclusion of this process was the total replacement of mail with plate, and this change was completed by the early years of the fifteenth century. The 'harness' or homogeneous suit of plate armour which by its nature had to be made to fit its wearer, the knight or mounted man-at-arms, and which covered him entirely (except for armpits and bottom) with plates of metal, was fully developed by about 1410–20 and remained in use for over two hundred years. During this period the essential features of its design remained constant, although there were some changes in the details of its form and decoration, and considerable changes in the material from which it was made.

FIFTEENTH-CENTURY PLATE ARMOUR

This was a product for the wealthy, made to order, and sold at a high price. The most important centre for the manufacture of arms was northern Italy. From the preceding century, English, French and German nobles bought imported Milanese armour. This trade has been described by Origo[13] from contemporary documents.

Such armour was generally made of steel and frequently hardened by heat treatment (see Figure 1). It was sold as being proof against arrows and crossbow bolts, and, even allowing for some exaggeration by the salesman, this was the expected level of performance.

Milanese armour was marked with the initials of the craftsmen who made it, and sometimes also with the marks of the men who sold it; initially perhaps also craftsmen, but later merchants, like the Missaglia. It was suggested by Buttin[14] some years ago that the presence of a double mark (see Figure 2) was the indication of 'double proof' – that is, armour that was resistant to more powerful weapons and which had been tested by the heavy crossbow rather than the sword, but the author has not been able to find any correlation between the metallurgy of the armour and the presence of double marks (see Table 1; note that that all tables appear at the end of the paper).

After the middle of the fifteenth century, the armourers of south Germany began to rival those of north Italy. The most important centres of production were Augsburg, Nuremberg, Innsbruck and Landshut. Their products were marked with a city mark to show that they satisfied guild requirements, but whether this meant that they were all proofed is debatable. They were marked also, especially in the case of fine armour, with a craftsman's mark as well.

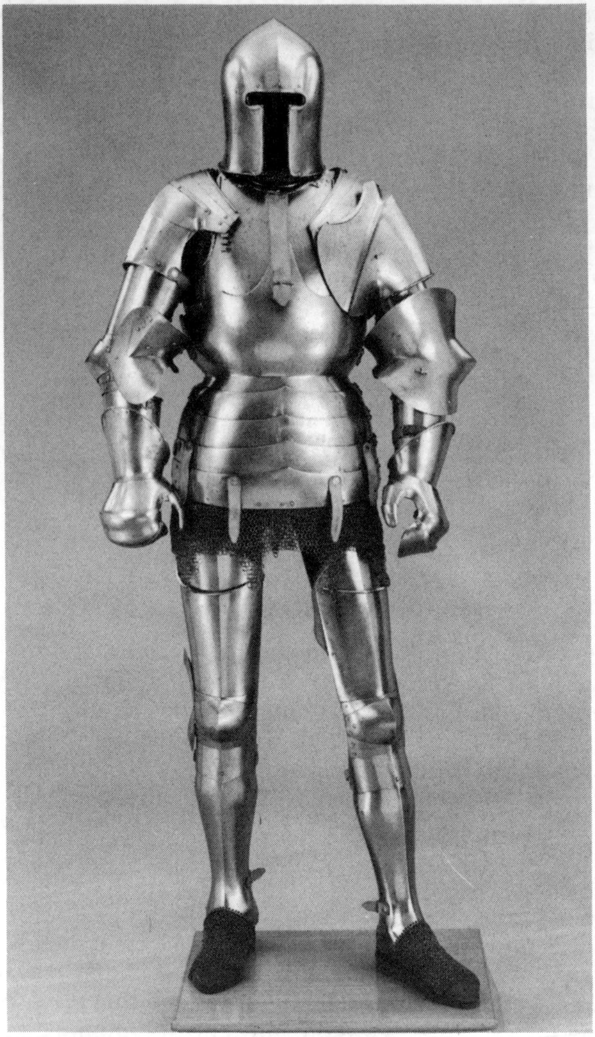

Figure 1 (i) The oldest complete armour in Britain. A high-quality armour made in Milan around 1440–50 for one of the lords of Matsch. It bears a number of makers' marks, some of which have been identified as those of craftsmen who specialized in making arm- or leg-defences. It was purchased from Churburg Castle, the home of the Matsch and Trapp families, in 1934 by the collector Hearst and later acquired by R.L. Scott (for £7,000) and it is now in the City Museum and Art Gallery, Glasgow (Inventory Number 39–65e). The curved surfaces deflected arrows, and the lance rest which would have been attached to the right side of the breastplate enabled a rider to support the impact of his lance with a rigid breastplate rather than with just his own wrist. It weighed 28 kg without mail skirt or gussets. Seven out of the ten plates from this armour examined by the author were made of hardened steels; two were unhardened steels; and one was iron. The hardness varied between around 200 and 300 VPH.
Photograph courtesy of Glasgow Museums.

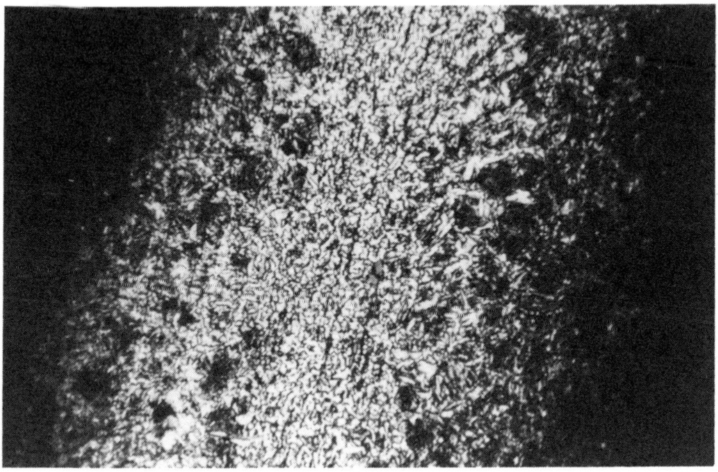

Figure 1 (ii) Photomicrograph of Figure 1 (i) magnification X 80. A cross section of the breastplate, which shows a steel of varying carbon content, hardened by heat treatment. The light areas are ferrite grains, while the black areas are nodular pearlite. This is typical of a slack-quenched steel. Note the lower-carbon band in the centre.
Photomicrograph by the author.

The earliest surviving German armour (see Figure 3 (i)) is from *c.* 1450 and is in the museum of the City of Vienna, but lacks a city mark. Little armour is marked as Augsburg before about 1480, so it can be identified as 'German' only on stylistic grounds (see Table 2). These 11 specimens from Augsburg armours (of 1480 onwards) are all made of hardened steel, of average hardness 338 kg mm^{-2}. But only 14 out of 20 specimens from Innsbruck armours of the same period were made of hardened steel (see Table 3). The average hardness of all these specimens from Innsbruck was 373 kg mm^{-2} (units of the Vickers Pyramid Hardness scale).

There is a general increase in hardness as fully quenched steels replace partly quenched ones. (See pp.127–8 for an explanation of these terms.) This does not seem to have occurred as quickly in Innsbruck as in Augsburg; indeed, the process was a difficult one to master. It may well be that the necessary techniques were introduced to Innsbruck by craftsmen from Augsburg. It is notable that in the sixteenth century, full-quenching and tempering was the procedure of choice in hardening Innsbruck and Augsburg armour of quality (and was successfully combined with etching and gilding to decorate it) until almost the end of the century.

Most infantrymen endeavoured to wear some armour. Although they could not readily bear the same weight as a horseman, a market for mass-produced cheap ('munition') armour was soon developed (see Figure 4 (i)). At the same time, the increasing threat from

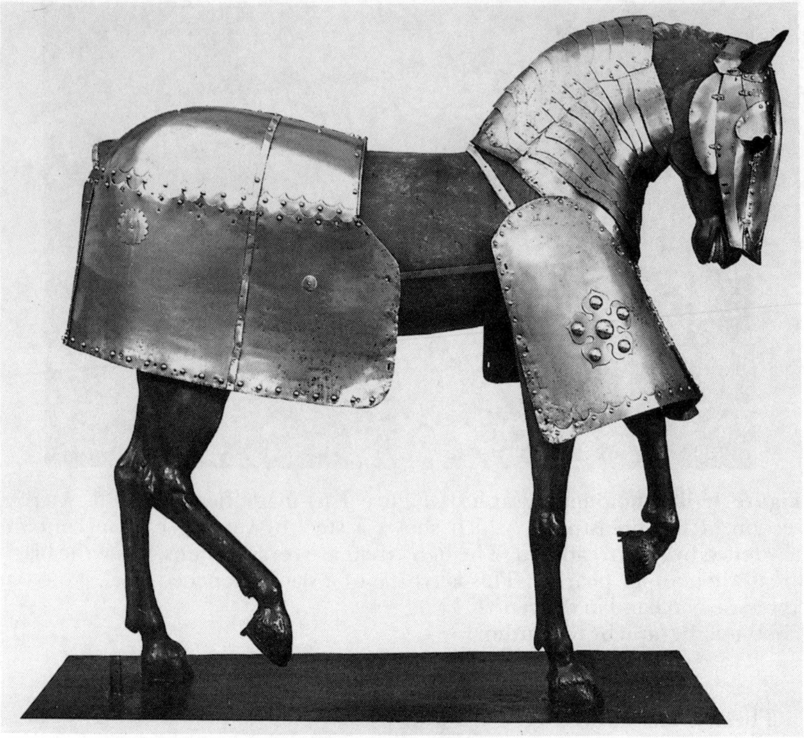

Figure 2 (i) The oldest surviving horse armour. An armour for a horse, made in Milan around 1450, and bearing the maker's mark of Pier Innocenzo da Faerno. It is now in the Museum of the City of Vienna (Inventory Number 127153). Notwithstanding its size (perhaps the largest steel object made before the nineteenth century), attempts were made to harden it by heat treatment. This was examined in six different places. All were low-carbon steels; 3 of the 6 plates had been quench-hardened, but the carbon content was generally too low (0.1–0.2%C) for any massive increase in hardness to result. The average hardnesses were *chanfron* (head defence) centre 150, left side 236, top 200 VPH; *peytral* (chest defence) top right side 252, lower right side 200, middle left side 220 VPH.
Photograph courtesy of Historisches Museum der Stadt Wien.

firearms meant that armour of quality had to protect its wearer against more powerful weapons. German armourers responded to this in the closing years of the fifteenth century by developing techniques to harden the steel by heat treatment. But the Milanese seem to have given up this struggle after about 1510, and concentrated on the decorative aspects of their products.

The use of armour did not start to decline until the last quarter of the sixteenth century, when the infantry were the first to abandon it. After 1600, musketeers generally wore no armour, while pikemen might wear a cuirass (breast- and back-plate), tassets (thigh defences)

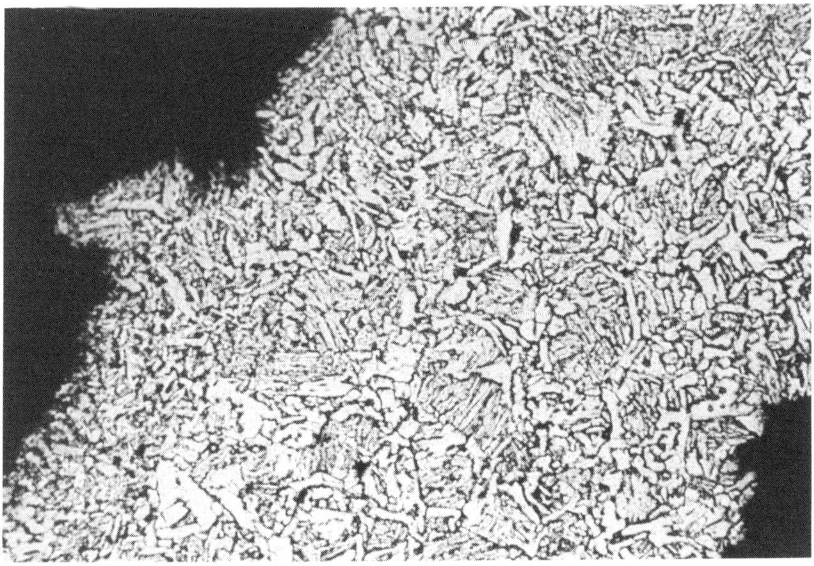

Figure 2 (ii) Photomicrograph of Figure 2 (i) magnification X 200. This sample is from the left plate (middle area) of the *peytral* (chest defence). The microstructure contains ferrite (light areas) and low-carbon martensite. The average hardness was 220 VPH.
Photomicrograph by the author.

and pot helmet only. Cavalrymen continued to fight in armour 'three-quarters' (i.e. to the knee) in the English Civil War and the Thirty Years' War to around 1640. Further east it was worn for longer: Austrian cavalry fighting the Turks continued to wear some armour, and the last active armourer in Europe was one Anna Maria Nierndorfer in Graz (fl. 1685).[15]

IMPROVEMENTS IN FIREARMS

It seems obvious that, once the gun had been invented, large guns should soon have found employment in siege warfare. The hitherto impregnable Roman walls of Constantinople were battered by the giant bronze cannon of the Turks in 1453. The French army captured sixty castles from the English in a year in 1449–50.[16] What is far from obvious is the reason for the equally rapid adoption of the handgun. The disadvantages of hand firearms were numerous. While such weapons were cheap enough, gunpowder was expensive. They were slow to reload, susceptible to interference by rain and erratic in performance, but they offered compensating advantages. The first written mention of them is in a Florentine MS of 1326,[17] and the

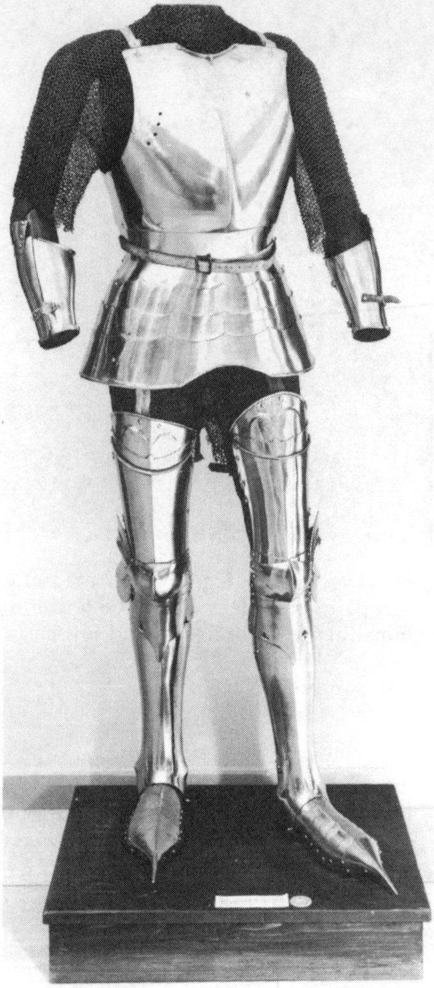

Figure 3 (i) The oldest surviving armour made in Germany. This does not bear a maker's mark, so that it can only be said to have been made around the middle of the fifteenth century – probably in south Germany. Its form is more angular than those made in Italy, and the metal is of inferior hardness and much less homogeneous. The armour is now in the Museum of the City of Vienna (Inventory Number 127000). Samples were taken from 6 different places within this armour; 3 were iron, 1 was a low-carbon steel, 2 were hardened but low-carbon steels, namely: breastplate, right side, iron 206; breastplate, left side, low-carbon steel 180; breastplate, plate below, iron 225; backplate, iron 155; arm, quenched low-carbon steel, 200; leg, quenched low-carbon steel, 328; The hardness of the breastplate was 180 VPH, but that of the leg reached 328 VPH.
Photograph courtesy of Historisches Museum der Stadt Wien.

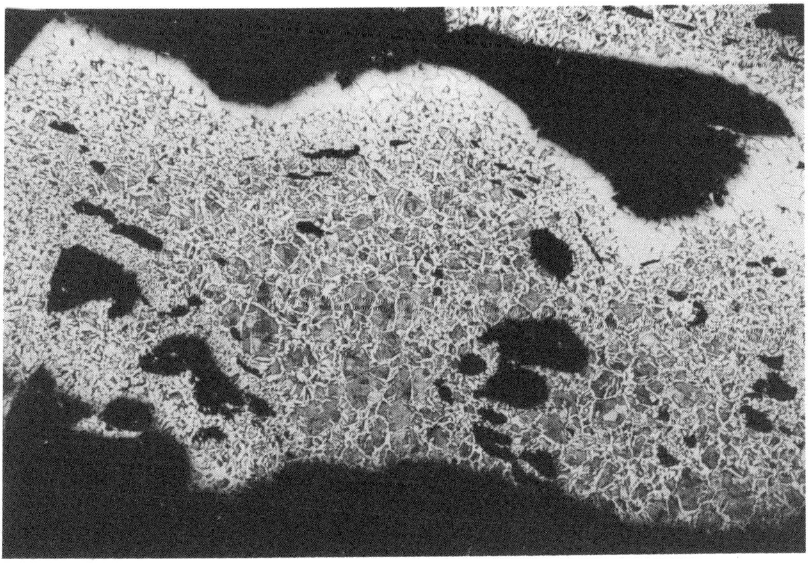

Figure 3 (ii) Photomicrograph of Figure 3 (i): magnification × 40. This sample is from within the breastplate. The irregular dark areas are pearlite and the white gains ferrite. This is typical of an air-cooled steel. Photomicrograph by the author.

earliest surviving example of one is the late-fourteenth-century cast bronze handgun from Moerkoe now in the National Historical Museum, Stockholm. Within a century such expensive novelties had come into widespread use, and the army of John the Fearless, Duke of Burgundy, included at least 4,000 handgunners in 1411.[18] In the Hussite wars (1419–34) the use of handguns helped the Czech peasants to defeat the German knights. They were particularly suited to employment by men who could not afford the time to train in archery skills. The militia of Milan was rearmed with handguns in 1449[19] and the Turkish janissaries started to discard their bows for guns after 1475.[20]

The performance of the handgun was steadily improved throughout the fifteenth and early sixteenth centuries by increasing the attainable muzzle velocity. The first handguns had been scaled replicas of artillery pieces, short and squat in their proportions. Increasing the charge of gunpowder in such pieces would have had little effect on the velocity of the bullet as it would have left the barrel before combustion was completed. Lengthening the barrel, however, enabled all the powder to be burned and the bullet to be accelerated for a longer period and therefore leave the muzzle with a higher velocity. Accordingly, from the early fifteenth to the mid-sixteenth century the average calibre (ratio of barrel length to bore) of

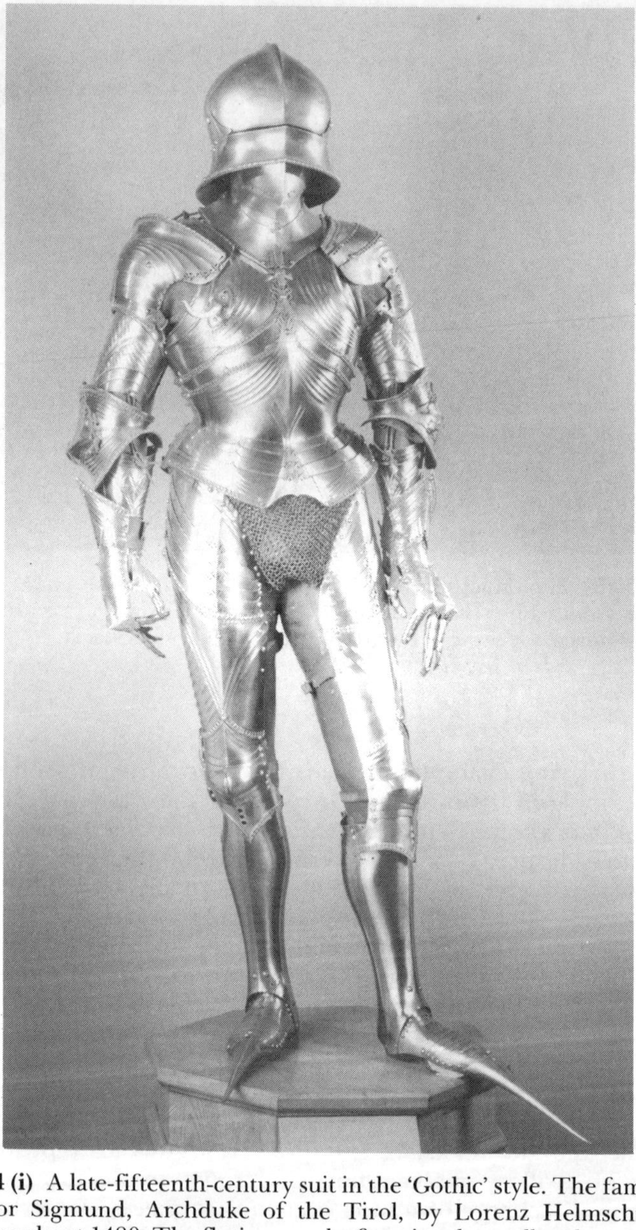

Figure 4 (i) A late-fifteenth-century suit in the 'Gothic' style. The famous suit made for Sigmund, Archduke of the Tirol, by Lorenz Helmschmied of Augsburg about 1480. The fluting may be functional as well as decorative, but the long toes of the foot-defences merely replicate late fifteenth century civilian fashions. It is now in the collection of the Kunsthistorisches Museum, Vienna (Inventory Number A.62).
Photograph courtesty of the Kunsthistorisches Museum, Vienna.

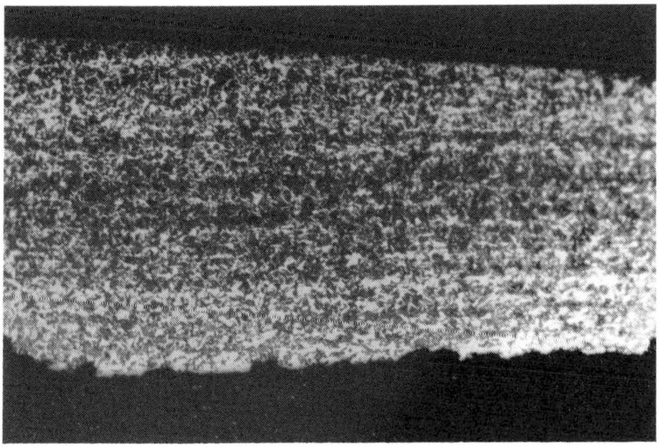

Figure 4 (ii) Photomicrograph of Figure 4 (i): magnification × 40. Cross-section of an associated gauntlet A62b. The microstructure is banded, containing grains of ferrite (white) and martensite (dark). This is a steel of heterogeneous carbon content. It has been quenched and tempered to harden it, but only those parts full of carbon have been transformed by this treatment. Average hardness 312 VPH.
Photomicrograph by the author.

handguns steadily increased (as the table of data from Sixl[21] shows) and, with it, muzzle velocity and armour-piercing ability. When the barrel became so long (about 75 cm) that both hands were needed to hold it, the 'serpentine' or pivoted trigger holding a length of smouldering cord was devised to fire it. The trigger could be pulled by the fingers of one hand while the gun was being held to the shoulder and aimed by both hands; this strictly speaking, was the 'arquebus'. When the barrel became so long (about 120 cm) that even two hands were insufficient to steady it, the gun had to be supported by a rest, a forked stick thrust into the ground. This was the 'musket', which was used to such notable effect by the Spanish infantry at the battle of Pavia in 1525. Pescara's musketeers 'frequently shot through both knights and their horses . . . lead balls were now shot by heavier handguns'.[22]

The other factor was the gunpowder used. When the three ingredients were mixed, they were moistened and ground together. This was the procedure followed in the fourteenth century, initially perhaps to avoid frictional heating. If, however, the moistened mixture was forced through a series of graduated sieves prior to drying, the gunpowder would be obtained in a uniform (and selected) grain size.[23] Such powder was said to be 'corned' and was more powerful than uncorned or 'serpentine' powder because the speed of burning of every grain was exactly the same. Fine-grain corned powder was generally used for muskets, and slower-burning coarse-grain corned powder for artillery.

The cost of a handgun was less than that of a steel crossbow. More important, the cost of gunpowder, although considerable, fell during the fifteenth century, and remained steady during the sixteenth century (a period of high inflation, and thus representing a fall in real terms) owing to the mass production of saltpetre in nitre-beds. A handgun which might cost 2*s* in 1446 had grown into an arquebus by 1511, and a musket costing 27*s* by 1558. A steel crossbow cost 6*s* 8*d* in 1482. The cost of a pound of gunpowder supplied to the English crown fell from 18*d* in 1346 to 8*d* by 1449 and 6*d* by 1515. Even after a century of inflation, corned powder was still only 7*d* a pound in 1599.[24]

None of this improved accuracy. The inaccuracy of a smooth-bore gun is due to the ball's wobbling in the barrel before leaving it and lengthening the barrel would do little to rectify this.

Some experimental results obtained by the author illustrate the effect of these developments (shown in Table A).[25] The author used 20 g of uncorned powder to fire both lead and iron bullets from a barrel of variable length at a sheet of mild steel 2.5 mm thick.

Table A

Barrel	*Calibre*	*Muzzle velocity* ($m\,s^{-1}$)	*Distance* (*m*)	*Penetrations*
254 × 19	13	149	10	6 out of 20
381 × 19	20	239	10	7 out of 11
914 × 19	48	255	20	7 out of 13
1372 × 19	72	343	20	8 out of 13

Note: Calibre here means the ratio of barrel length to bore.

Some results that have been obtained by others in the past were generally comparable. Grancsay shot balls from a wheel-lock musket, arrows from a bow of 30 kg draw weight, and bolts from a crossbow of 330 kg draw weight against various specimens of armour from his private collection at 5 m range. When an arrow or bolt struck the armour a glancing blow, it was deflected; if it struck it straight, it pierced it. Two out of three bullets penetrated the helmet, and two out of three the breastplate, but none out of three penetrated a sixteenth century backplate of unspecified but evidently superior quality. This illustrates the advantages that the curved surfaces of plate armour would offer compared with mail; it would be extremely difficult to hit them at a perpendicular angle. His missiles were timed and the bullets travelled at 190 $m\,s^{-1}$, the arrows at 41 $m\,s^{-1}$ and the bolts at 43 $m\,s^{-1}$.[26]

Lasson constructed a replica handgun of about nine calibre (200 × 23 mm), loaded it with 38 g of uncorned powder and a 50 g lead ball, and fired it from 30 m when it penetrated a 'suit of light armour' of unspecified quality.[27]

The first widespread use of handguns seems to have been in the Hussite wars of the Bohemian peasant revolutionaries against their (largely German) feudal lords. The dimensions of a number of Hussite guns from Plzen have been published recently by Fryda.[28] Most of these are in the range of 30 to 40 calibre, so their muzzle velocity would be around 200–250 m s^{-1}.

DEVELOPMENTS IN ARMOUR: ITALIAN ARMOUR IN THE SIXTEENTH CENTURY

These improvements in the muzzle velocity of handguns made them such threatening weapons to armoured knights that their first reaction was fear. The Chevalier du Bayard, who fought with the French armies in Italy from the battle of Fornovo (1495) until his death by a musket shot in 1524, and famous for his chivalry in other contexts, had all captured handgunners hanged. His contemporary, the *condottiere* Gian Paolo Vitelli, had them blinded instead.

For reasons which are not entirely clear, the metallurgical quality of the best Italian armour seems to have deteriorated quite suddenly in the early sixteenth century. Armour made for noblemen and decorated with patterns of etching and gilding was now made of iron or low-carbon steel. Twenty-five specimens from 16 armours dating from 1500 to 1610 in the Royal Armoury in Turin were examined by the author and only one of these specimens was found to be made of hardened steel.

This decline in the metallurgical quality of Italian armour is not paralleled in the armour of south Germany which continued, until the end of the sixteenth century, to be made of hardened steel to resist the growing threat of firearms.

One explanation of this anomaly might have been that two grades of armour were available — hardened or decorated — and while German armourers could offer both combined, the Italians could only offer a choice. But unfortunately this explanation must be discarded, because even plain Italian armour for battle does not seem to have been hardened.

Italian armour in the sixteenth century was still a popular, if expensive, item of military equipment — or perhaps rather of decorative metalwork. In the later part of the sixteenth century, the workshop of Pompeo della Cesa in Milan produced a large number of armours for noble and princely patrons such as Renato Borromeo and Wolf Dietrich von Raitenau, Archbishop of Salzburg. Pompeo used a better metal than his Milanese predecessors, as metallography shows, and this metal might have been made by a different process, as the analysis of slag inclusions suggests, but he was still unable or unwilling to harden the steel. It might be objected that these were purely parade armours, but they seem to have been expected to possess some utility in defence, for otherwise they could just as well have been made of iron (which they were not). Examples, from various museums which include those quoted from the Royal Armoury, Turin, and the Waffensammlung, Vienna, show that Pompeo generally used steel, and as will be discussed below (see 'Slag inclusion analysis', pp. 125–6) this was steel apparently made by a different process (see Table 5).

THE MANUFACTURE OF ARMOUR IN INNSBRUCK

At Innsbruck, capital of the Tirol and key to the Brenner Pass, there had been a flourishing colony of armourers established in the suburb of Muhlau since the middle of the fifteenth century.

Thomas and Gamber have traced the history of the Innsbruck armourers from an edict of Duke Sigmund in 1446, and the earliest armourers whose works (at first, much influenced by the north Italian masters) are known to us: Konrad Treytz the Elder, Konrad Vetter; and by the third quarter of the fifteenth century, their sons Jorg Treytz, Hans Vetterlein, together with Christian Schreiner (who married the daughter of Konrad Treytz the Elder), Christian Spor and Kaspar Rieder. They were succeeded in turn by Hans Schral, Christian Tretyz and his brother Adrian the Younger (who married Hans Prunner's daughter), and others who were encouraged to come from different parts of Germany, Hans Laubermann (called Kölner), Hans Rabeiler (called Pair = Bayer) and, most significantly, from Augsburg came Hans Prunner in 1482, and the brothers Jorg and Klaus Wagner soon after. These foreigners settled and prospered and their civic honours are recorded in the archives of Innsbruck and the Tirol.

When Maximilian succeeded to the empire in 1493 his ambitions exceeded those of his predecessor; he wanted his court armoury to be a production centre for munition armour as well as a workshop for making small numbers of valuable presentation armours. Conrad Seusenhofer was engaged as 'Hofplattner' in 1504 with responsibilities in both areas. When his contract was renewed in 1509 he was to be paid 200 gulden a year and would work only for the emperor; under him there would be six journeymen, four polishers and two apprentices. Maximilian commissioned an autobiographical work, the 'White King' (*Weiskunig*) in which it was said, 'the young king [i.e. Maximilian] had established in his city, Innsbruck, a great armoury in which he had had made various armours of new form . . . there were some of the family Treizsaurbayn that had made harness of so hard a kind that no one was able to pierce it with a crossbow [but] the knowledge was lost with their death. But there was a master residing at Muhlau, named Caspar Riederer, who had known these people as a boy, and passed on this knowledge to the young White King. After that, the king had the same art meditated upon [improved] and the Hofplattner, Conrad Seusenhofer, used it.'

By 1507, Seusenhofer had completed the work of organizing the court armoury and set up a polishing mill on the Sill to save a journey to Muhlau, in which suburb most of the armourers had previously resided. He arranged for supplies of sheetmetal (*Harnischblech*) from Leoben: each month 8 Säm (one Säm = one mule-load of about 20 pieces of sheet), and yearly 3 Säm steel and 2 Säm of a special steel called '*zwischach*' and perhaps corresponding to the 'double shear' steel of Sheffield. 'Shear steel' was made in Sheffield by carburizing Swedish bar iron, then folding this and forging it out into a bar. 'Double shear' was folded twice to improve its homogeneity.

At Innerberg one Hans Zanger made patterns (*matrizen*) for the mass production of munition armour. When very large quantities of armour were ordered, Seusenhofer enlisted the help of armourers from Muhlau to help fulfil the needs of the Arsenal.

Craftsmen like Konrad Treytz were well able to exploit the metal of the Erzberg (the very rich iron ore deposits near Leoben, work to this day), so one wonders why craftsmen were induced to settle from other parts of Germany — the reputation of Augsburg must have been remarkable indeed! What was it based upon? Metallographic study of armour from Augsburg and Innsbruck may suggest an answer (see Table 6).

The average hardness of 22 specimens from Innsbruck was 465 kg mm^{-2}, which is somewhat harder than in the fifteenth century.

The earliest piece with an Augsburg city mark that was hardened was a '*rennhut*' of 1486, while the earliest hardened Nuremberg piece (so far examined) was an armour of the Pfalzgraf Philip from 1525 (see Figure 5 (i). Both Augsburg and Nuremberg commonly produced hardened armour throughout the sixteenth century (see Table 7).

The average hardness of 29 specimens from Augusburg is 300 kg mm^{-2}, which is markedly less hard than the average of contemporary Innsbruck armours. So one wonders why Maximilian had imported craftsmen from Augsburg. It seems that their techniques of hardening steel were better, although they would not have had as good a raw material in Augsburg as they were to find in Innsbruck.

If the products ascribed to different members of one family are compared, then the results are an interesting illustration of this. Lorenz Helmschmied, active in Augsburg in 1490 (d. 1515), was the most celebrated armourer of his day.[29] His workshop was continued by Kolman, his son (d. 1532), and then Desiderius, his grandson (active to 1553). As one would have expected, the family used similar techniques for hardening their armous (quenching and tempering the steel), but they applied them to different raw materials. Lorenz Helmschmied and his son Kolman used steels of a banded microstructure and then hardened them to around 350 kg mm^{-2}, but those ascribed to Desiderius are more inconsistent, or perhaps less securely attributed. Some are of surprisingly poor quality, while others are decidedly better than the products of his grandfather.

Lorenz's reputation may have been based on his ability to produce

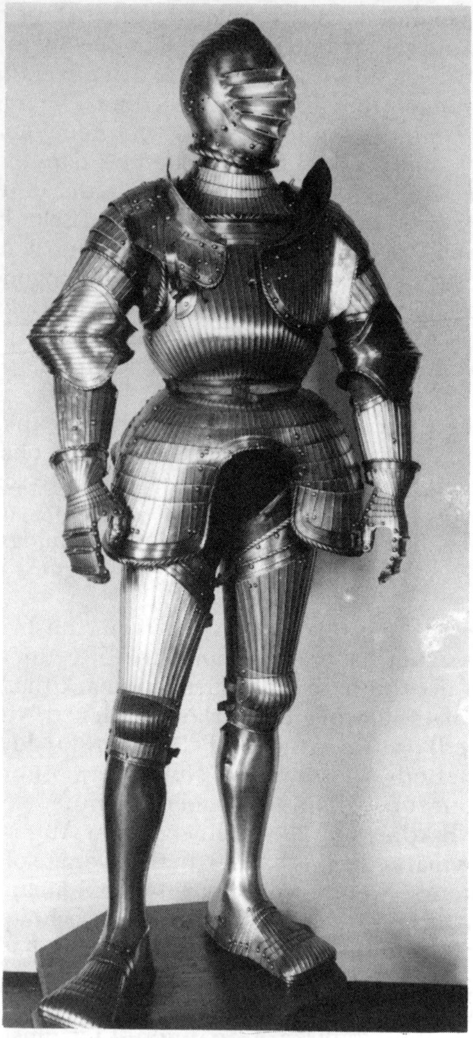

Figure 5 (i) A high-quality sixteenth century armour in the 'Maximilian' style. This style, with its surfaces covered with fluting, owes its name to its supposed invention by the Emperor Maximilian, although there is no evidence for this other than his noted patronage of armourers. It has been suggested that the fluting helped to stiffen the plates, but an equally likely reason could have been that they were thought useful in deflecting missiles, and the influence of fashion cannot be discounted. This particular example was made in Nuremberg by Wilhem von Worms for the Pfalzgraf (Elector Palatine) Philip the Warlike about 1525. A generation later, Ferdinand, Archduke of the Tirol, started collecting armour as mementoes of his famous contemporaries. This was to form the nucleus of the modern collections of the Kunsthistorisches Museum in Vienna, and Ambras (where this armour remains; Inventory no.WA.238)
Photograph courtesy of the Kunsthistorisches Museum, Vienna.

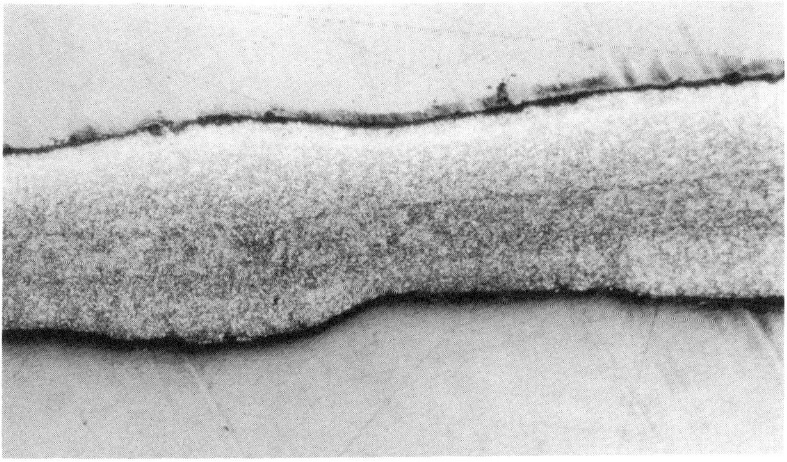

Figure 5 (ii) Photomicrograph of Figure 5 (i): magnification × 20. A cross-section of the helmet's skull. The microstructure is a uniform martensite. This is a quenched and tempered steel of hardness 488 VPH. The thin line of slag inclusions down the centre of the section may be the result of a fold during forging.
Photomicrograph by the author.

harder armour than previous Augsburg craftsmen, but since their work cannot be identified by either city or master's marks, this must remain an open question (see 6 (i) and 6 (ii)). Certainly his armours (identified from the Thun sketchbook) are notably less hard than the products of his Innsbruck contemporaries. Of course, the hardness of a steel depends on its carbon content as well as on its heat treatment. They are fairly consistent in composition, being made of banded steels, but the steels used by Lorenz were certainly lower in carbon than those available to the Innsbruck masters. His grandson made armours for the Kings of Spain (such as A420 shown in Figure 7 (i)), and no doubt had access to a wider range of raw materials.

Perhaps this is the explanation of the reputation of Augsburg craftsmen: they had mastered the techniques of controlling their hardness of steels by fully quenching and subsequently tempering them. Such were essential for making letter-punches for the printing industry, but any connection must be speculative. But the best material then available to them was the banded steel commonly found in medieval artefacts. This would have been the outcome of forging together either several pieces of carburized iron, or of piling and welding an unevenly carburized bloom. This would yield a microstructure of characteristically streaky appearance.[30]

Innsbruck craftsmen were blessed with a much better, more homogeneous raw material, but were less sure of the techniques needed to heat-treat it, and so slack-quenching was more frequently

Figure 6 (i) A typical infantry armour of the type mass-produced in Nuremberg in the sixteenth century. This is a munition armour of simple design and low cost made at many cities in Germany and the Netherlands without very much significant change in design throughout the sixteenth and early seventeenth centuries. Swiss mercenaries (and German *landsknechte*) were customers for this type of armour, and this is one of several hundred such still preserved in the Old Arsenal at Solothurn, Switzerland (Inventory No. 104). The breast- and back-plates are fastened to one another by adjustable straps which pass over the shoulders and sit upon the large gorget (collar). This allows the relative positions of breast and back to be changed to suit wearers of different sizes. The shoulder defences consist of a series of overlapping plates attached to the collar. Some had longer arm-defences ('splints') and were called 'Almain rivets' in England because they were also adjustable for size with sliding rivets. The total weight would generally be between 8 and 13 kg.
Photograph by the author.

employed in the middle of the fifteenth century. With the advent of Augsburg craftsmen and their intermarriage with Innsbruck families, trade secrets were shared, and Innsbruck armour was not only made of 'better metal' (an English gentleman of 1537, Sir R. Wingfield, writing to Thomas Cromwell and describing an Innsbruck armour as 'a complete harneeyse which was made for mysyllfe at Insbroke in

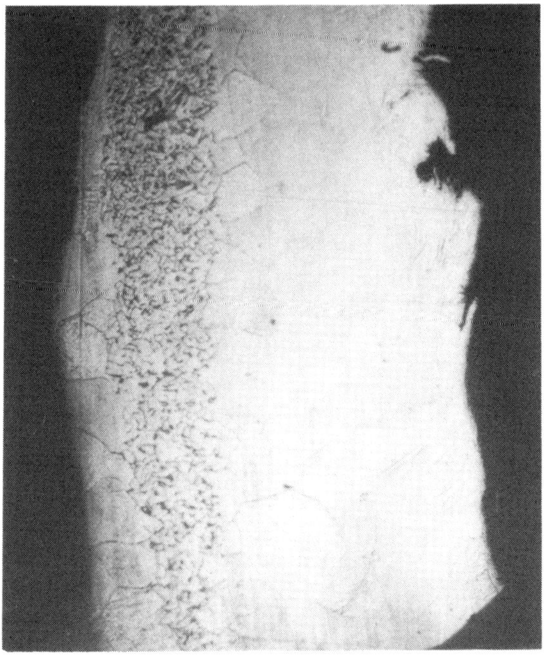

Figure 6 (ii) Photomicrograph of Figure 6 (i): magnification × 100. Most such armours are made of wrought iron or low-carbon steel. This typical microstructure contains mostly (white) grains of ferrite together with a band of higher carbon content where (dark) areas of pearlite are to be found. Photomicrograph by the author.

Austryk and gyven unto me by the Emperor Maximylan . . . I do warrant that a fayrer or *of bettyr mettell* can nott be found'[31]), but now hardened more effectively, so that early sixteenth-century Innsbruck armour scales a metallurgical peak as well as an artistic one.

Jakob Topf seems to have been the last Innsbruck armourer to employ this technique of quenching and tempering, and even then not for tournament armours. (Anton Peffenhauser was the last Augsburg armourer to harden his steels.) Jakob's son Hans Jakob Topf did not bother to do so, apparently believing that it was easier to make a field harness of 'proof' simply by using thicker plate, instead of by increasing its hardness.

THE MASS PRODUCTION OF ARMOUR

From the first years of Henry VIII's reign, arms and armour were regularly imported to England from Germany, Italy and Flanders. In 1509 Louis de Fava and Leonardo Frescobaldi (both Italians) were selling large supplies of armour to the king. In 1511 Henry

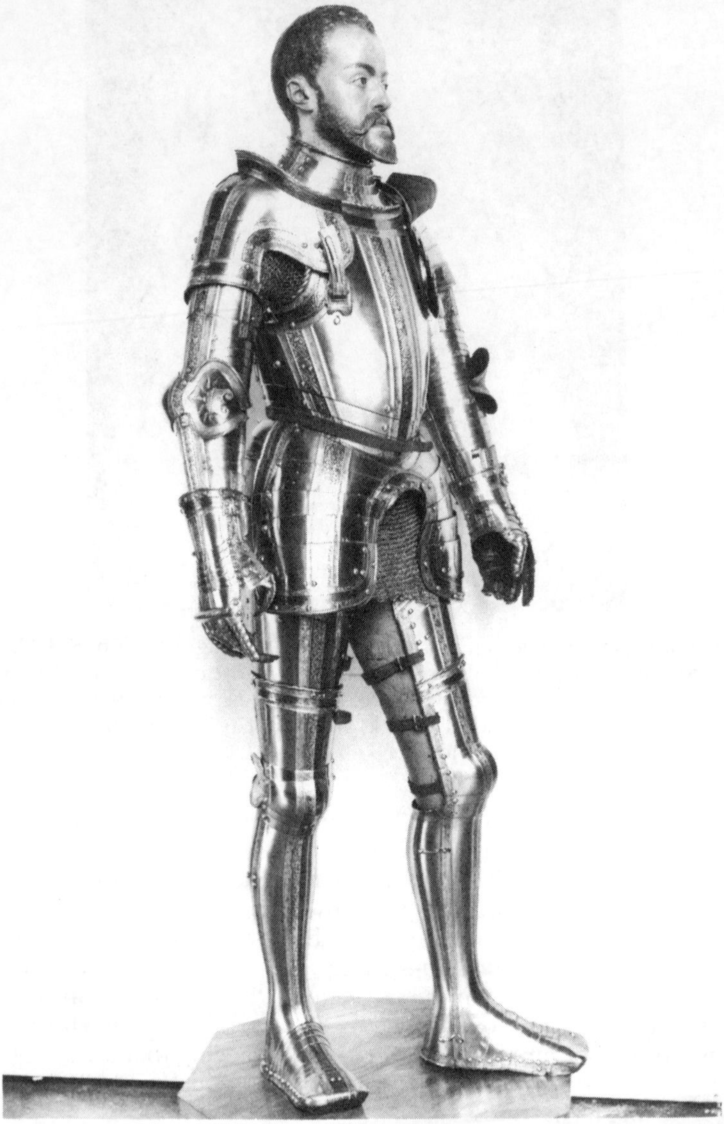

Figure 7 (i) An armour made for a king. This is the field armour from a garniture (set of interchangeable parts) made for King Philip II of Spain in Augsburg by Desiderius Helmschmied (grandson of Lorenz) about 1544. It is decorated with bands of etched and gilded patterns, so the heat treatment needed for hardening must have been combined with fire-gilding. It is now in the collection of the Kunsthistorisches Museum, Vienna (Inventory Number A.547)
Photograph courtesy of the Kunsthistorisches Museum, Vienna.

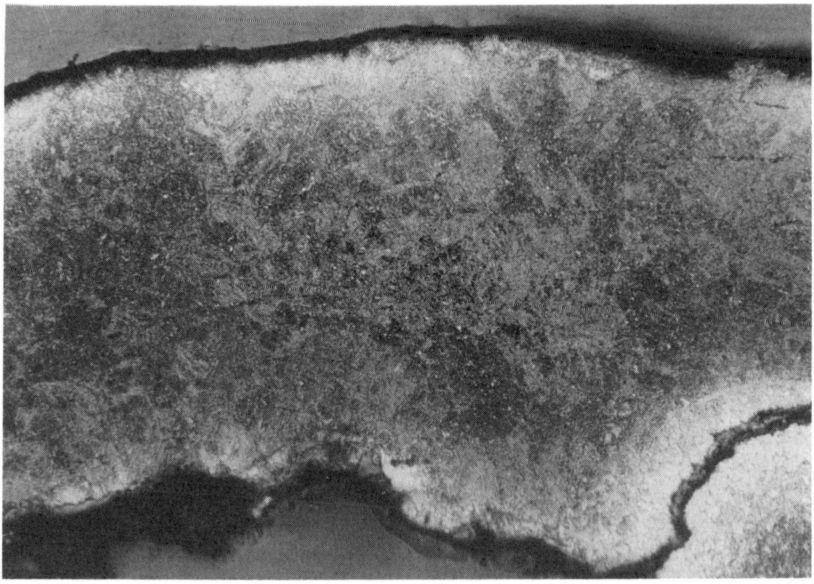

Figure 7 (ii) Photomicrograph of Figure 7 (i): magnification × 160. This is from the cross-section of the right gauntlet. The microstructure is uniform tempered martensite except for areas of ferrite associated with a crack full of grey solid; this is probably the remains of a faulty forge-weld which has locally decarburized the steel. Otherwise the section visible is quite homogeneous. The average hardness is 441 VPH.
Photomicrograph by the author.

despatched Richard Jerningham and two other gentlemen of his court to Germany and Italy to buy arms and armour. Some of this was for his own personal use; but in addition, Jerningham bought in Milan in 1513 5,000 foot-soldiers' suits, or 'Almain rivets'. At about the same time Henry had bought from a Florentine merchant, Guido da Portenari, 2,000 'Almain rivets', which were infantry armours consisting of sallet (open helmet), collar, breast- and back-plates and tassets (thigh-plates), for 16*s* the set. The price had risen since 1509, when 8*s*, and in 1512, 11*s* had been paid for similar suits of armour. In 1539, by contrast, King Henry looked to northern Europe and bought 1,200 'complete harness' (horsemen's armours, with protection for the arms and legs) at Cologne for £454 (carriage paid) and 2,700 at Antwerp for £630,[32] so the cost had fallen to about 7*s* 6*d* or less. Since the material must have cost at least 3*s* and workmen were paid around 9*d* a day, the cost of making the armour must have been between two and six days' wages, not counting fuel or overheads.

In 1560, Sir Thomas Gresham, Queen Elizabeth's agent in Flanders, exported 8,000 corselets (infantry armours) from Antwerp, and then bought another 6,000 which he was forced to ship

Figure 8 (i) One of the last fine armours made in Europe. This armour was made in the English Royal Workshops at Greenwich about 1590 for Thomas Sackville, Lord Buckhurst, who would have had to pay for permission to have an armour made there for him. It was originally heavily blued and decorated with bands of etching and gilding, much of which survives today. It is now in the Wallace Collection (Inventory No. A62). An extra breastplate and other pieces for use in tournaments were made at the same time and decorated in the same way to form a garniture. A sample was also taken from within the breastplate; its microstructure was of all-tempered martensite, and its hardness was 325 VPH.
Photograph courtesy of the Wallace Collection, London.

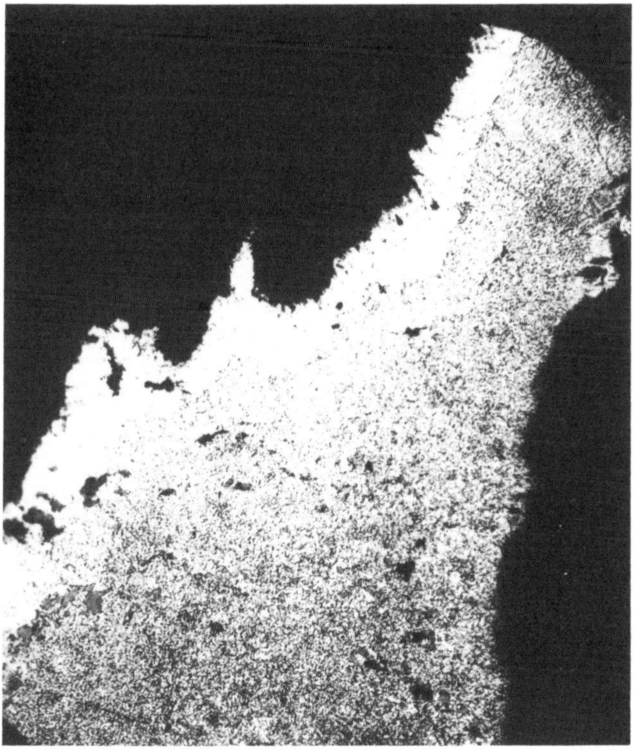

Figure 8 (ii) Photomicrograph of Figure 8 (i): magnification × 80. This is a cross-section of the left (lower) arm-defence. This is a banded steel; layers of ferrite (white) and martensite (dark) are visible. The hardness is 281 VPH. Photomicrograph by the author.

surreptitiously via Hamburg, spending altogether £12,000 on these and other munitions.[33]

As long ago as 1468 a suit might cost £7 (around 100 days' wages), so evidently techniques of mass production were being employed for some armour. In 1542, a mandate to the Lord Mayor of London fixed the price of a horseman's armour at £2 5*s*.[34] Later in that century the Armourers' Company of London petitioned against the importation of foreign armour and offered to supply Queen Elizabeth with up to 8,000 armours over five years at a cost of £3 6*s* 8*s* for a complete horseman's armour and 30*s* for a corselet.[35]

By contrast, an armour made at Greenwich (and very similar to that shown in Figure 8 (i)) in 1612 for Prince Frederich Ulrich of Brunswick, with extra pieces for the tournament, and elaborate decoration by etching and gilding, cost no less than £400 – almost the same price differential as that between a Lada and a Jaguar XJ220! Allowing for careful selection of steel, its successful hardening by heat treatment, more careful fit of the armour and fine decoration, one may still ask: Was the metal made in the same way? Slag inclusion analysis may supply us with a partial answer.

FURNACES

Iron ore was charged into a furnace with charcoal, and in some cases with limestone as well.[37] Charcoal is burned in front of the tuyeres and a stream of hot gases (CO, CO_2, N_2) ascends while the ore descends. At high temperatures, the ratio of CO to CO_2 is high, which enables metal oxides to be reduced. FeO reduces readily at about 800°C, MnO at about 1400°C, but the other oxides present are unlikely to be reduced at the temperatures available in small (about 1 m in height) shaft furnaces; SiO_2 needs about 1600°C to reduce.[38] The rate of reduction will depend on the particle size and the porosity of the ore, and the fuel/ore ratio, which in turn determines the concentration of CO and the temperature. Under the most favourable conditions, iron might be formed in a furnace of very modest dimensions—a bowl hearth in fact.[39] Otherwise iron particles will start to form at some point on their journey down the shaft. Solid iron thus formed will absorb carbon from the hot, CO—rich gases until it reaches the tuyere zone. If it absorbs a significant amount, its melting point will fall, perhaps even as far as the ambient temperature in the furnace, in which case it will melt, and then dissolve more carbon very quickly from direct contact to form the eutectic 'cast iron'.[40] The unreduced oxides present from the ore as well as from the clay and stones of the furnace lining (CaO, Al_2O_3, SiO_2, and probably MnO) as well as any unreduced FeO, will react together to form a slag, a glass-like material the free-running temperature of which will depend on its composition: 2FeO. SiO_2 (fayalite), 1205°C; FeO. CaO. SiO_2, 1220°; CaO. SiO_2, 1540° (close to the melting point of pure iron).

In the simplest form of bloomery, a 'bowl hearth', perhaps less than 1 m high, the iron might be reduced, but neither the iron nor the slag melted. The products would then have to be crudely separated by breaking them apart, or else reheating them at a higher temperature to melt away the slag. A larger furnace, with a shaft around 2 metres high might be operated as a 'bloomery hearth' to give as products a 'bloom' of solid iron and an iron-rich slag liquid at around 1200°C. On the other hand, if the shaft furnace was high enough, it might be operated as a 'blast furnace' to give liquid iron rich in carbon (and perhaps manganese or even silicon as well) and a low-iron slag liquid at a much higher temperature.

Any bloomery iron will contain entrapped slag inclusions of similar compositions to the tap-slag. If a manganese-rich ore was used then MnO will still be in the slag. In a 'blast furnace' liquid iron will separate readily from liquid slag, so that the metal produced will be free from tap-slag. This liquid iron was at first regarded as a useless by-product, then used as a cheaper substitute for bronze in casting. Large cannon were being cast in iron by 1390, and at some later stage methods for converting it to forgeable iron were discovered.[41]

If an 'indirect' or two-stage process was employed, the liquid cast iron would have been converted to iron or even steel by being wholly or partially decarburized in a 'finery' by melting it and then submitting the liquid to an oxidizing atmosphere, or allowing it to drip through an air blast onto a charcoal hearth.[42]

Some iron oxide would form as the carbon content of the iron was reduced and might react with the lining of the hearth (SiO_2 and perhaps Al_2O_3 and CaO could be present in the lining of the hearth, depending on its construction) so that the wrought iron produced in the finery would contain some finery slag, which might differ little in composition from extraction slag. Published analyses of slags from fineries show that most of the elements present in bloomery slags are also present in finery slags, so that the chemical composition alone is not likely to be a satisfactory way to distinguish bloomery from finery slags.[43]

The blooms of iron formed in the finery could be large (the eighteenth-century, but essentially similar, 'puddling' process produced blooms of 50–70 kg) and needed hammering into sheets or bars with a water-powered hammer, another medieval development. Much of the finery slag would be squeezed out during this, but much would remain and a certain amount of smithing slag (FeO and SiO_2) would also be incorporated, dependent on the amount of working. Of course, bloomery iron would contain smithing slag also, but not finery slag. Direct (bloomery) iron might then differ from indirect (blast furnace plus finery) iron as follows:

- Direct: metal would not contain Mn or Si; inclusions would consist of furnace slag and smithing slag.
- Indirect: metal might contain Mn and perhaps Si; inclusions would consist of finery slag and smithing slag.

Steel remained a luxury product throughout the Middle Ages. According to Thorold Rogers, the price at which raw iron was sold in England varied between 0.45p per pound (*c.* 1300), 0.84p per pound (*c.* 1400), and 0.44p per pound (*c.* 1500); it was usually sold by the hundredweight (50 kg). On the other hand, steel was sold at 1.65p per pound (*c.* 1300), 1.60p per pound (*c.* 1400) and 1.20p per pound (*c.* 1500); it was sold by the piece, later by the sheaf, garb, fagot, cake or barrel.[44] The words sheaf, garb, and fagot all have a similar meaning — that of a bundle, whether of sticks, arrows, or pieces of metal. The price difference suggests that steel was made separately from iron, and with three or four times as much difficulty. It might have been made in one of several ways.

(1) Directly, in a shaft furnace with operating conditions midway between those of a bloomery and a blast furnace. The slag

inclusions present would resemble those from bloomery iron, but would be lower in FeO, since conditions would be more reducing. Traces of manganese or titanium oxides in the ores would not be reduced, and so would remain in the slag.

(2) By case-carburizing pieces of bloomery iron, or selecting higher-carbon fragments from a heterogeneous bloom after breaking up, and forging them together (this might be the connection with those names derived from bundles). The slag inclusions present would be unchanged but mixed with more smithing slag.

(3) The decarburization of liquid cast iron might be halted at an intermediate carbon content, that of steel. According to Walzel ([45]) steel was made in Styria this way by letting the liquid iron from the blast furnace drip through an air blast onto a charcoal hearth. Obtaining anything like a consistent carbon content would have been difficult, if not impossible, and British attempts to make 'puddled steel' by a similar direct process in the nineteenth century proved to be unsatisfactory.[46] Likewise, Bessemer found that it was easier to remove all the carbon and then add a measured weight to give steel of the chosen carbon content.[47] A steel made this way would contain finery and smithing slags.

(4) A method related to (3), sometimes called the 'Brescian process', was described by Biringuccio in 1540 (to be precise, he ascribed it to Valcamonica, near Brescia) and copied by Agricola a few years later.[48] A lump of bloomery iron ('weighing thirty to forty pounds') was to be swirled about on the end of an iron rod in a bath of liquid cast iron for 4 to 6 hours, with crushed marble added, until it was somewhat carburized, and then taken out and forged into a uniform product. This would contain a mixture of bloomery, finery and smithing slags. So, the different methods might produce steels with:

	(1)	(2)	(3)	(4)
Metal inclusions	*Mn (perhaps)*	*No Mn, no Si*	*Mn, Si (perhaps)*	*Mn, Si (perhaps)*
Furnace slag	Yes	Yes	No	Yes
Smithing slag	Yes	Yes	Yes	Yes
Finery slag	No	No	Yes	Yes

Unfortunately for this neat classification, it is not always possible to distinguish between the different types of slag inclusions. It has been suggested that the absence of MnO in slag indicates a smithing rather than an extraction slag.[49] It is also possible that MnO might be reduced in a blast furnace, instead of being left in the slag, but then reoxidized in the finery, so that the presence of MnO cannot really be used to make deductions about the furnace. The absence of MnO might well indicate a smithing slag, formed by the sprinkling of sand on to iron being forged. On the other hand, its absence might just mean that an ore low in manganese was used in the first place. So, while the analysis of elements present in slag inclusions may not give unambiguous information about the iron- or steel-making processes, detecting the presence of different types of inclusion may be much more useful, since each type should be the result of a different operation. The proportions of the different types may indicate the history of the armour's manufacture.

SLAG INCLUSION ANALYSIS

With this prospect in mind, the author set about having the slag inclusions in seventy samples of armour examined in a scanning electron microscope by means of energy-dispersive X-ray diffraction. The detailed results have been published elsewhere;[50] only the conclusions need concern us now. Out of 600 inclusions 566 fell into one of four categories:

Type 1 Iron oxide (Fe only), perhaps hammer scale from forging or a corrosion product.

Type 2 Iron silicate (Fe and Si only), with the approximate proportions of fayalite (70% FeO), which is assumed to be a smithing slag since no other metals are present.

Type 3 Mixed metal oxides and silicates (Fe, Si, and small amounts of Ca, Al, and Mn) assumed to be a furnace slag, since their proportions approximate to those of analysed bloomery slags already published.

Type 4 Mixed metals oxides and silicates (Fe, Si, Al and Ca, but without Mn); this might be a finery slag, but is also possibly an extraction slag from a low-Mn ore.

The preponderant type was 3, which made up 433 out of 600. If the assumption that this was a bloomery slag was correct, then this was hardly a surprising result. To take the results of examining armour of quality first: twenty-five out of twenty-eight samples from German armours contained only or mostly Type 3 inclusions. Out of thirty-one samples from Italian armours, twenty-six contained only or mostly Type 3 inclusions. But among the exceptions, a group stands out; armours from the workshops of Pompeo della Cesa contain both Type 2 and Type 3 inclusions (and only the products of this workshop do this). Otherwise there seemed to be no significant variation with date or place of manufacture. It may be deduced that, *for fine armour, iron or steel made in a bloomery was employed*. The cost of even the best steel would have been only a small proportion of the total cost, and so there would have been little incentive for the armourer to seek a cheaper source for his steel. The exception to this, Pompeo, was producing armours of steel (hardenable, but unhardened) in the last quarter of the sixteenth century, long after other Italian armourers had reverted to using wrought iron even for their most costly armour. Pompeo may have wanted to use a stronger material without the difficulties of heat-treating a metal that was going to be gilded (a problem which the Germans, but not the Italians, seem to have overcome). Pompeo's steel could have come from both a bloomery and a finery since it contains different types of slag, and it is an attractive hypothesis that the process of mixing cast and wrought iron described by Biringuccio was in fact the process employed by Pompeo's suppliers.

When samples of munition armour were examined, much more variation was found. Of the five Italian infantry armours examined, all had Type 3 but two had numerous Type 2 inclusions as well, i.e. a mixture of types like that shown in Pompeo's products. It may be evidence of finery metal, or it may be that cheaper armour underwent more smithing, so it was more likely to contain metal from more than one source.

Five examples of German munition armour were examined, and all had inclusion populations *totally different* from those found in fine armours. Type 2 and Type 4 inclusions predominated, although they differed from one another. Their matrices wre also found to have high P contents and, in some cases, high S as well. This is probably due to the use of high-phosphorus ores and perhaps the use of coal in smithing as well.

While there might be little point in seeking cheap metal for fine armour, the cost of the metal was a significant fraction of the cost of munition armour, so there would have been every incentive to exploit novel techniques of ironmaking if they yielded cheaper iron. There is not enough evidence to link the finery positively with munition armour, but one can say that the metal for some munition armour was made by methods which were very different to those used to make the metal for fine armour.

THE ABANDONMENT OF ARMOUR

So advances in metallurgy had made armour plentifully available. The government of Elizabeth made strenuous efforts to ensure that the militia was better armed and armoured to resist Spanish invasion. Elizabeth exhorted the Lords Lieutenant in each county to call musters of the militia more frequently and to see that they were fully attended by properly equipped men, to train the shire-levies regularly and to equip them (at county, not Crown, expense) with muskets and ammunition and, where necessary, with armour.[51] There was a lively controversy over its usefulness, involving Sir John Smythe and Sir Richard Hawkins among others.[52]

The opening of the Civil War in England nevertheless saw the importation of much armour from Holland.[53] Armour was still widely used in the Thirty Years' War and by the Austrians against the Turks,[54] but by the middle of the seventeenth century it was falling out of use. It was feasible to make knightly armour that was bulletproof (or at least pistolproof[55]), but this was armour of high quality which could not be mass-produced. In order to render it bulletproof, mass-produced (munition) armour could only be made thicker, and therefore more difficult for the infantry to wear. There was a general belief (which might well have been erroneous) that armour had to be bulletproof to be worth wearing. Maurice de Saxe was to argue strongly against this in the middle of the eighteenth century, observing that 'the effects of gunpowder in engagements are become less dreadful, and fewer lives lost by it, than is generally imagined. I have seen whole volleys fired without even killing four men.' He went on to describe how, for instance, at the battle of Belgrade (1717) only thirty-two Turkish cavalry had been brought down by a combined volley from two battalions of Austrian infantry, and added later, 'I am at a loss to know why armour has been laid aside, for nothing is either so useful or ornamental . . . it was the fashion in Henri IV's reign and since, and powder was introduced long before that time . . . its disuse was occasioned by nothing more than the inconvenience of it.'[56] But by the time he was writing this, the military argument was well lost and armour had become the province of antiquarians, leaving behind only its legacy of the blast furnace.

Acknowledgements
The author would like to thank the Armourers & Brasiers Company, the British Academy and the Society of Antiquaries for their financial help. He would also like to record his thanks to the many museum curators and their staffs whose co-operation has made this study possible.

TABLES OF RESULTS

The metallurgy has been deliberately simplified in the tables by calling: IRON all examples of wrought iron and low-carbon steel (up to around 0.3% C); STEEL all examples of medium-carbon steels (around 0.5% C) which have not been hardened by quenching; HARD all examples of steel which have been hardened by either full- or slack-quenching (these terms are explained below), except that low-carbon steels which may not have been much altered by quenching are classed with IRON (on the grounds that if an armourer had quenched a steel with insufficient carbon content, there would have been little improvement in performance).

Although steel is harder and stronger than iron, the principal advantage to be gained from employing it is the dramatic increase in hardness obtained when the steel is quenched, i.e. plunged red-hot into cold water.

This increase is the result of the formation of a new, different crystalline form in the steel (pure iron is unaffected by quenching). If a steel is allowed to cool in air after forging, then equilibrium conditions will prevail. The carbon which was dissolved in the iron above 900°C comes out of solution as a lamellar arrangement of iron carbide and ferrite called *pearlite* which has a distinctive microscopical appearance.

If however the steel is cooled more rapidly (i.e. by quenching) and equilibrium is not attained, then other crystalline products may form, namely:

(1) pearlite in a nodular form, almost irresolvable, called in the older literature *troostite*;
(2) a material of acicular appearance harder than pearlite, called *bainite*;
(3) very rapid cooling may form *martensite*, a material of lath-like appearance and great hardness.

Much depends on the carbon content and the dimensions of the object being treated, but quenching in water generally results in such rapid cooling that an all-martensite structure is obtained (this is called 'full-quenching'). Quenching in oil, molten lead, or some other coolent less drastic than water may produce a mixture of micro-constituents — pearlite and bainite was well as martensite — as may an interrupted or delayed quench (such procedures are collectively called 'slack-quenching' and nowadays avoided). The customary modern procedure for hardening a plain carbon steel is to fully quench it, and then to reheat it carefully to 'temper' it.

Tempering reduces the hardness of martensite somewhat, but by removing most of the internal stresses, it reduces its brittleness and hence increases its impact strength. Successful tempering requires a steel of consistent composition and accurate control of temperature and time. These requirements would have been difficult to meet in the Middle Ages, and so it is not surprising that slack-quenching seems to have been the preferred heat treatment of the Milanese armourers, presumably on the grounds that only one operation was involved, and although the product would have been of inferior strength to a fully quenched and tempered steel, there was less danger of embrittlement on the one hand or overtempering (and thus softening the steel) on the other. But many of the Augsburg and Innsbruck armourers seem to have employed a two-stage process for heat treatment as the microstructures of their hardened products show only martensite, or tempered martensite, sometimes with ferrite but without pearlite or bainite.

The results are grouped in columns as follows:

(1) Approximate date of manufacture.
(2) Museum and inventory number.
(3) Metallurgy (iron, steel or hardened steel).
(4) Microconstituents.
(5) Vickers Pyramid Hardness (kg.mm. $^{-2}$).
(6) Journal of publication.

The museums in which these armours are located are indicated as follows:

Ch = Churburg Castle Collection, Sluderno, Italy.
WC = Wallace Collection, London.
TLA = Royal Armouries, Tower of London.
WSV = Waffensammlung, Vienna.
MSM = Stadt Museum, Munich.
VSM = Stadt Museum, Vienna.
BNM = Bavarian National Museum, Munich.
G = City Museum and Art Gallery, Glasgow.
S = Stibbert Collection, Florence.
Tor = Royal Armoury, Turin.
MMA = Metropolitan Museum of Art, New York.
Amb = Waffensammlung Schloss Ambras.
Stut = Württembergisches Landesmuseum, Stuggart.
VC = Waffensammlung Veste Coburg.
Fitz = Fitzwilliam Museum, Cambridge.
DHM = Deutsches Historisches Museum.
Dres = Historical Museum, Dresden.
K = City Museum, Cologne.
SA = National Museum of St Angelo, Rome.

Details of the microconstituents are also given in the tables: F = ferrite, P = pearlite, vfP = very fine pearlite, B = bainite, C = carbides, M = martensite.

Fuller details, including photomicrogaphs, have been published by the author elsewhere (in the papers signified by a, a2, etc.).

a = 'Fifteenth century armour from Churburg,' *Armi Antiche* (1986), 32:3.

a2 = 'Italian armour in the Royal Armoury of Turin', *Armi Antiche* (1987), 33:27.

h8 = 'A note on the armour of Henry VIII', *Archaeologia*, 1979, 106:157.

j = 'Four helms of the 14th century compared', *Journal of the Arms & Armour Society* (1981), 10:80.

j2 = 'Italian armour and Cosimo dei' Medici', *ibid.* (1991), 13: 293.

m = 'A note on the manufacture of armor in 15th century Italy', *Metropolitan Museum Journal* (1979), 13:131.

mz = 'Die Herstellung von Rüstungen im 15 und 16 Jahrhundert', in *Das Münchner Zeughaus*, ed. R.H. Wackernagel (Munich, 1983), 47.

w = 'The manufacture of armour in Germany: the metallurgical evidence', *Waffen- und Kostumkunde* (1987), 29:90.

w2 = 'Metallography and the manufacture of armour in Innsbruck', *Waffen- und Kostumkunde* (1990), 32: 53.

* = This publication.

In Tables 1 to 8 (except 4), column headings are as follows:

Column I = Museum and Date.
Column II = Inventory or catalogue number.
Column III = Metal used.
Column IV = Microstructure of sample.
Column V = Hardness of sample.
Column VI = Place of publication with further details.

Table 1 Fifteenth century Italian armour.

I		*II*	*III*	*IV*	*V*	*VI*
1390	Ch	13	iron	F+P	180	a
1390	TLA	IV.470	steel	P		a
1400	BNM	W 195	hard	P+B+M	340	j
1400	WC	A.252	steel	P		a
1400	WC	A.69	iron	F+P		a
1400	TLA	IV.497	steel	P		a
1410	Ch	16	iron	F+P	200	a
1430	MMA	29.158.5	iron	F+C	184	m
1430	TLA	III.1123	steel	P		a
1440	MMA	42.50.2	hard	P+B+M	338	m
1440	TLA	III.1093	hard	P+B+M	340	a2
1440	TLA	IV.430	hard	M+B	400	a
1440	TLA	IV.5	iron	F+C		a
1445	G	39.65e	hard	M+B+F	(several)	a
1450	Ch	19	steel	P		a
1450	Ch	21	hard	F+B+M	340	a
1450	TLA	IV.498	hard	F+M		a
1460	TLA	II.168s	hard	F+M		a
1470	Ch	67	hard	F+M	280	a
1470	TLA	IV.424	iron	F+P		a
1475	Ch	23	steel	F+P	250	a
1475	Ch	33	hard	F+M	300	a
1480	TLA	II.168g	hard	F+M		a
1480	ST	3880	hard	M	520	j2
1480	WSV	A.3	hard	F+C	294	j2
1480	Tor	C65	iron	F+P		a2
1490	Tor	B19	iron	F+P		a2
1490	WSV	B.2	hard	M	464	j2
1490	WSV	A.111	iron	F+C	187	j2
1495	WSV	A.5	hard	M	600	j2

I	*II*		*III*	*IV*	*V*	*VI*
1495	WSV	B.35	iron	F+P	254	j2
1500	Ch	56	hard	M	690	a
1510	Ch	70	iron	F+P		a
1510	Ch	69	iron	F+P		a
1510	WSV	A.11	hard	M	400	j2
1510	ST	3122	iron	F+P	200	j2
1510	ST	3146	iron	F		j2

Table 2 Fifteenth century Augsburg armour.

I	*II*		*III*	*IV*	*V*	*VI*
1450	VSM	z76	steel	F+P+C	225	*
1480	WSV	A60	hard	F+M	275	w2
1480	VSM	z77	hard	TM	442	w2
1485	MSM	1353j	hard	TM	345	mz
1485	WSV	A62b	hard	F+M	312	w2
1485	WSV	A79	hard	F+M	379	w2
1486	WSV	A79k	hard	F+M	352	w2
1486	VC	c52	hard	F+M	315	w
1490	VSM	126	hard	F+M	302	w2
1490	Chur	42	hard	F+B	351	a
1495	WSV	A7	hard	F+M	310	w2

Table 3 Fifteenth century Innsbruck armour.

I	*II*		*III*	*IV*	*V*	*VI*
1460	Ch	27	steel	F+P+C		a
1460	Ch	24R	hard	F+P+M	339	a
1470	TLA VI.	375	hard	F+M		w2
1475	Ch	22	hard	F+M	289	a
1480	TLA	II.1	iron	F+P		a
1480	TLA	III.1284	steel	vfP		j2
1480	Ch	24	hard	TM+F	520	a
1480	Ch	24S	hard	TM+F	410	a
1480	Ch	29	steel	vfP	226	a
1485	Ch	49	hard	P+B	346	a
1485	Ch	51	hard	M+B+F	439	a

I	II		III	IV	V	VI
1485	MSM	1353c	hard	vfP+M+B	405	mz
1489	WSV	A9	hard	TM	451	w2
1490	Ch	31	hard	F+P+M	308	a
1490	TLA	III.2562	hard	TM+F		j2
1490	Ch	41	hard	F+B	245	a
1490	MSM	z837	hard	F+vfP	311	w2
1490	Ch	62	hard	F+C	224	a
1495	Ch	47R	hard	M	868	a
1495	Ch	47L	iron	F	224	a

Table 4 Dimensions of some Hussite guns in the City Museum of Plzen (Pilsen).

Bore (mm)	*Barrel length*	*Calibre*
35	315	9
26	290	11
29	415	14
22	626	28.5
21	943	45
21	847	40
21	916	44
24	675	28
36	1070	30
26	998	38
26	860	33
24	935	39
24	976	41
22	900	41
20	747	37
16	643	40
20	853	43
22	853	39
20	600	30
22	710	32
24	820	34
22	974	44
24	925	38.5
24	904	38

Table 5 Italian armour of the sixteenth century (1510 and after).

I	*II*		*III*	*IV*	*V*	*VI*
1510	Tor	B19	iron	F+P		a2
1530	Tor	E50	iron	F+P	210	a2
1530	SA	945	steel	F+P	234	j2
1540	Tor	C11	iron	f		a2
1540	WSV	A 406	iron	F	180	j2
1560	WSV	A 118	iron	F	210	j2
1560	Tor	B7	iron			a2
1560	Tor	C10	steel	F+P		a2
1565	Tor	B4	steel	F+P	228	a2
1570	Tor	B8	iron	F		a2
1570	Tor	C15	hard	vfP	400	a2
1580	TLA	II146	iron	F+P		j2
1580	Tor	B10	steel	P	260	a2
1585	St	3476	steel	F+P	226	j2
1590	BNM	1011	iron	F+P	238	w
1590	St	2561	iron	F		j2
1590	Tor	C21	steel	F+P		a2
1590	Tor	B3	steel	F+P		a2
1590	Tor	C70	steel	F+P	230	a2
1600	Tor	B35	iron	F		a2
1602	Tor	C98	iron	F		a2
1602	Tor	E18	iron	F+P		a2
1610	Tor	E21	iron	F+P		a2

Table 6 Sixteenth-century Innsbruck armour.

I	*II*		*III*	*IV*	*V*	*VI*
1500	TLA	IV.468	hard	P+M		h8
1510	Fitz	109h	hard	TM	352	w2
1510	TLA	III1700	hard	TM		w2
1510	Ch	71	hard	F+M	333	a
1510	Ch	72	hard	TM	485	a
1510	Ch	80	hard	P	293	a
1511	WSV	A186	iron	F+C	180	w2
1511	WSV	A244	hard	M	468	w2
1514	WSV	A109	iron	F+C	276	w2
1514	WSV	A179	hard	TM	401	w2

I	II		III	IV	V	VI
1515	MSM	1355a	hard	F+P+C	255	mz
1520	St	3465	hard	F+P+M	310	
1525	VSM	z43	hard	M+P	293	w2
1530	VSM	z46	hard	TM+P	514	w2
1530	VSM	136	hard	M+F	517	w2
1535	VSM	z94	hard	M+F	468	w2
1535	Ch	98	hard	F+M	392	a
1537	VC	c53	hard	vfP+F+M	325	w
1537	BNM	W648	hard	TM+F	<520	h8
1540	TLA	II 23	hard	M	249	w2
1540	Ch	169h	hard	F+M		w2
1549	TLA	II 169	hard	F+P+M	283	w2
1550	BNM	W647	hard	F+C+M	570	w2
1550	Stut	6530	hard	TM		w2
1550	Amb	331	hard	TM	498	w2
1555	Amb	767	iron	F	134	w2
1555	St	2891	iron	F		w2
1560	BNM	1479	hard	M	395	w2
1570	WSV	1496	hard	F+M	259	w2
1580	Amb	3010	steel	F+P	191	w2
1580	Amb	1568	iron	F	149	w2
1580	BNM	1510	hard	TM	505	w2
1582	WSV	1113	hard	TM	415	w2
1619	WSV	1534	steel	F+P	263	w2
1641	WSV	1704	steel	P+F	298	w2

Notes Z.Cat. refers to the catalogue 'Das Wiener Burgerliche Zeughaus' (1977), and the z-numbers to this catalogue. Many pieces have also (different) new and old inventory numbers. Also, < = up to.

Table 7 Sixteenth century Augsburg armour.

I	II		III	IV	V	VI
1508	VSM	z54	hard	M	265	w2
1510	WSV	A310	hard	F+M	331	w2
1520	VSM	z93	hard	F+C+M	470	w2
1520	MDG	2323	hard	F+B		w2
1524	WSV	A374	steel	P+F	281	w2
1525	TLA	III851	steel	vfP+F		w2

I	*II*		*III*	*IV*	*V*	*VI*
1525	Tor	c5	hard	F+P+C	316	w2
1525	TLA	III771	steel	vfP		w2
1540	Tor	c69	iron	F+P	115	w2
1543	WSV	A546	steel	F+P	240	w2
1544	WSV	A547	hard	TM	441	w2
1548	WSV	A420	hard	TM	407	w2
1549	WSV	A610	steel	P+F	305	w2
1550	VSM	z634	steel	F+C	233	w2
1550	St	c135	hard	TM	377	w2
1550	BNM	4752	steel	F+P	252	w
1551	Tor	c15	iron	F+P		a2
1555	St	2822	iron	F+P	230	
1558	Amb	B144	iron	F+P	179	
1580	VC	55	hard	M+B	435	w
1580	VC	chan	iron	F+P	129	w
1580	BNM	14/L	hard	F+P+M	329	w2
1582	Dres	M98	hard	TM		w
1588	Dres	M28	hard	TM+F+P		w
1588	Dres	M124	hard	TM		w
1588	Dres	M29	iron	F+P		w
1591	TLA II	186	hard	F+P+M	360	w2
1600	VC	I23	iron	F+P		w
1610	K	953	iron	F		w

Table 8 Armour made by members of the Helmschmied family

I	*II*		*IV*	*V*
Lorenz Helmschmied				
1480	WSV	A60	F+M	275
1485	WSV	A62b	F+M	312
1486	WSV	A79k	F+M	379
1492	WSV	A791	F+M	352
1495	WSV	A7	F+M	310
Kolman Helmschmied				
1510	WSV	A310	F+M	331
1520	VSM	93	F+C+M	470
1520	DHM	2323	F+B	

I		*II*	*IV*	*V*
1524	WSV	A374	P+F	281
1525	TLA	III851	vfP+F	
1525	Tor	c5, 6	F+P+C	316
Desiderius Helmschmied				
1538	TLA	III771	vfP	
1540	Tor	c69	F+P	115
1543	WSV	A546	F+P	240
1544	WSV	A547	TM	441
1548	WSV	A420	TM	407
1551	Tor	c15	F+P	
1560	Fitz	4h*	P+F	200

*This armour is only in the style of Desiderius Helmschmied.

Notes and References

1. M. Rusu, 'Das keltische Fürstengrab von Ciumesti in Rumänien', *Germania*, 50: 267.
2. V. Hruby, 'Ein Ringpanzer der Hallstattzeit', *Sbornik Praci Filosoficke Fakulty Brnenske Universite*, VIII 1959, 33.
3. Varro, *De lingua latina*, V.116.
4. H.R. Robinson, *The armour of Imperial Rome* (London, 1980), *passim*.
5. J.W. Anstee, *Museums Journal*, 53: 200.
6. J. Lang *et al.*, in D. Tweddle, *The Coppergate Helmet* (York, 1993).
7. This was made of wrought iron (private communication from Prof. Radomir Pleiner).
8. G. Arwidsson, *Valsgarde 8: Die Grabefunde von Valsgarde* (Uppsala, 1954).
9. A.N. Kirpicnikov, 'Russische Waffen des 9–15 Jahrhunderts', *Waffen- und Kostumkunde*, 1986, Parts 1 and 2, 1 and 85.
10. J.F.C. Fuller, *Armanent and History* (London, 1946), 73.
11. A.V.B. Norman, 'Notes on a newly-discovered piece of 14th century armour', *Journal of the Arms and Armour Society*, 1975, 8: 229.
12. B. Thordeman, *Armour from the Battle of Visby*, 2 vols (Stockholm, 1939), 167.
13. I. Origo, *The Merchant of Prato* (London, 1957), 36 and 355. L.G. Boccia and E.T. Coelho, *L'arte dell'armatura in Italia* (Milan, 1967), *passim*.
14. C. Buttin, 'Notes sur les armures à l'épreuve', *Revue Savoisienne* (Annecy, 1901) 2/3: 64.
15. Dr Hollister-Short points out that one Elizabeth Bennett was the chief blacksmith at Blenheim Palace under Sarah Churchill's orders, and a near contemporary of Anna Maria Nierndorfer.
16. C. Oman, *The Art of War in the Middle Ages* (London, 1924), *passim*.
17. J.R. Partington, *A History of Greek Fire and Gunpowder* (Cambridge, 1961), 99.
18. C. Brusten, *L'armée bourguignonne* (Bruxelles, 1953), 108.

19. J.C. Sismondi, *Histoire des républiques italiennes du moyen âge* (Paris, 1818), IX, 341.

20. D. Ayalon, *Gunpowder and Firearms in the Mamluk Kingdom* (London, 1956), 51.

21. P. Sixl, 'Entwicklung und Gebrauch der Handfeuerwaffen', *Zeitschrift fuer historische Waffenkunde* (Dresden, 1900–2), an article in a series of eight parts.

22. O. Delbrueck, *Geschichte der Kriegskünste* (Berlin, 1921), III, 58.

23. Partington, *op. cit.* (17), 324.

24. M.M. Postan (ed.) *Cambridge Economic History of Europe* (London, 1966), Vol. I, Ch. 8. A.R. Williams, 'The manufacture of saltpetre in the Middle Ages', *Ambix*, 1975, 22: 125.

25. A.R. Williams, 'Firing tests with a simulated 15th century handgun', *Journal of the Arms and Amour Society*, 1975, 112.

26. S.V. Grancsay, 'Just how good was armor?', *True Magazine*, April, 1954, 45.

27. T. Lasson, 'From hand-cannon to flintlock', *Gun Digest*, Chicago, 1956, 33.

28. F. Frýda, 'Plzeňská Městská Zbrojnice' (Pilsen, 1988).

29. B. Thomas, 'Lorenz Colman (recte Helmschmied), armourer of Augsburg' *Apollo Annual* (London, 1948), 21.

30. B. Gilmour and R.F. Tylecote, 'The metallography of early ferrous tools and edged weapons', *British Archaeological Reports*, 155, 1986: also R. Pleiner, 'Zur Technik von Messerklingen aus Haithabu', *Berichte über die Ausgrabungen in Haithabu*, 1983, 18:63; and *id.*, 'Die Technologie des Schmiedes in der Grossmährischen Kultur', *Slovenska Archeologia* 1967, 15:77.

31. F.H. Cripps-Day, *Fragmentata Armamentaria: an introduction to the study of Greenwich armour* (privately published, Frome, 1934), Vol. 1, Part 2, 41.

32. *Letters and Papers of King Henry VIII*, Part 2, Appendix 14.

33. J.W. Burgon, *Life of Sir Thomas Gresham*, 2 vols (London, 1839), Vol. I, 325.

34. Ibid, xviii, 412.

35. C.C. ffoulkes, *The Armourer and His Craft* (London, 1912), 184.

36. J.G. Mann, *Catalogue of an Exhibition of Armour Made in the Royal Workshops at Greenwich* (London, 1951), 54.

37. G.R. Morton and J. Wingrove, 'Constitution of bloomery slags, Part II: medieval', *Journal of the Iron and Steel Institute*, July 1972, 478.

38. For a useful summary see J.A. Charles, 'The coming of copper and iron', in *The Coming of the Age of Iron*, ed. T. Wertime (Yale, 1980), 151.

39. See, for example, R.F. Clough, 'The bloomery process', in *The Crafts of the Blacksmith*, ed. B. Scott (Belfast, 1984), 19.

40. R.F. Tylecote, J.N. Austin and A.E. Wraith, 'Mechanism of the bloomery process in shaft furnaces', *Journal of the Iron and Steel Institute*, 1971, 290: 342.

41. O. Johannsen, *Geschichte des Eisens* (Düsseldorf, 1953), 217; and also B. Awty, 'The origins of the blast furnace', *Historical Metallurgy*, 1987, 21: 96, and (1988) 22: 125.

42. R.F. Tylecote, *Metallurgy in Archaeology* (London, 1962), Ch. 10.

43. G. Sperl, 'Die Technologie der direkten Eisenherstellung', in *Erz und Eisen in der Grünen Mark* (Graz, 1984), 99.

44. J.E. Thorold Rogers, *A History of Agriculture and Prices in England, 1259–1793*, 7 vols (1866–1902). There are too many tables to cite individually, but all may be consulted with advantage.

45. R. Walzel, 'The 2000-year tradition of the Austrian iron and steel industry', *Journal of the Iron and Steel Institute*, August 1951, 366.

46. K.C. Barraclough, 'Puddled steel', *Journal of the Iron and Steel Institute*, 1971, 785, 952.

47. T. Wertime, *The Coming of the Age of Steel* (Chicago, 1962), 289.

48. V. Biringuccio, *Pirotechnia* (Venice, 1540), Eng. trans.: C.S. Smith and M.T. Gnudi, 1942 (reprinted New York, 1959), 67. G. Agricola *De re metallica* (Basel, 1556), Eng. trans.: H.C. Hoover and L.H. Hoover, 1912 (reprinted New York, 1945), 342.

49. J.G. McDonnell, 'The study of early iron smithing residues', in Scott, *op. cit.* (39), 47.

50. A.R. Williams 'Slag inclusions in armour', *Historical Metallurgy*, 1991, 24: 69.

51. L. Boynton, *The Elizabethan Militia 1558–1638* (London, 1967), 65; also C.G. Cruickshank, *Elizabeth's Army* (London, 1966), 109.

52. C. Blair, *European Armour* (London, 1958), 143.

53. T. Richardson, 'Russell Robinson's Dutch armour', *Journal of the Arms and Armour Society*, 1991, 13: 256.

54. P. Krenn, *Der Grazer Harnisch in der Türkenabwehr* (Graz, 1971), 8–20.

55. H.A. Dillon, 'A letter of Sir Henry Lee and the trial of iron for armour', *Archaeologia*, 1901, 51: 167.

56. Saxe, Maurice de, Marshal of France, *Reveries or Memories upon the art of War*, trans. W. Fawcett (London, 1757).

ICOHTEC: Some Informal Personal Reminiscences

MELVIN KRANZBERG

At the XIXth International Congress of the History of Science in Zaragoza, Spain, in August 1993, the International Committee for the History of Technology (ICOHTEC) celebrated its twenty-fifth anniversary with an International Symposium entitled 'The Place of Theory in the History of Technology.' Speakers came from the United Kingdom, United States, Russia, Poland, Germany, Finland, France and Australia – and scholars from other countries (e.g. Italy, Sweden and Japan) also took part in other Congress sessions dealing with technological history.

What's so remarkable about that? One expects scholars from different nations to participate in international meetings, and since ICOHTEC is a Scientific Section of IUHPS (International Union of the History and Philosophy of Science, which is part of UNESCO), its meetings should attract scholars throughout the world.

But when one recalls that 1993 marked the twenty-fifth anniversary of ICOHTEC, that gives one pause for thought. For 1968, the year of ICOHTEC's 'official birth', – and of the Prague Spring – was a time when the Cold War was at its height – or, to the 'thermometer-minded', at its lowest point if measured in terms of the deepening chilliness of the atmosphere between the Western nations and the Eastern European bloc, then dominated by the Soviet Union.

Hence the fact that historians of technology from both East and West could come together at that time to form ICOHTEC indicates one of the brighter sides – and hopes – for scholarly learning. Indeed, it shows how *philosophia*, the Greek word meaning 'love of knowledge' – in this case, the desire to learn about the role of technology in human history – can sometimes overcome political suspicions and ideological differences. But while advances in human technological prowess show how mankind can benefit from technical progress, history also demonstrates that technology interacts with the sociocultural context. Hence, conflicting economic goals, political rivalries and antagonistic social ideologies might narrow technology's potential benefits to mankind, and people might even use – or, rather, misuse – technology in a way that is harmful to the human species and our planet Earth.

Technology, one must remember, is a very human activity: a

product of the human mind, imagination and skills working together. While some people have used it to impose their will upon other human beings, it can also be employed to help one's fellow humans. And, as ICOHTEC's history demonstrates, the scholarly study of technological history is a wonderful tool for advancing human knowledge and friendship.

Just as history begins with pre-history, so the story of ICOHTEC began before it became organized and won 'official' status as a Scientific Section of IUHPS. That's not surprising, for just as the institution of marriage in most cultures involves a premarital period when bride and groom first meet, become further acquainted, and finally commit themselves to permanent ties, so ICOHTEC's 'wedding' of historians from East and West was preceded by a 'get-acquainted' period of courtship.

Their coming together seemed unlikely at the time, because of the Cold War. While their common interest in technological development throughout history was the major factor in bringing these historians together, it should also be remembered that the tremendous growth in the destructive power of weapons during World War II — embodied in the development of rocket-propelled missiles by German scientists and the role of the atom bomb in bringing the war to a conclusion — created fears that threatened the 'one-world' envisioned by the founders of the United Nations in the post-war period.

These fears were heightened when on 4 October 1957 the Soviet Union launched *Sputnik*, the first artificial satellite, and a few years later (12 April 1961) Yuri Gagarin became the first human being to orbit Earth in a space vehicle. But these major accomplishments, while signalling great advances in the human capacity to 'triumph' over nature, also engendered fear in the West — for they indicated that the Soviets possessed sufficient mastery over science and technology to destroy the Western nations should the Cold War ever 'boil over'.

The strength of those fears was indicated by the quick responses made by the United States government. Thus, following *Sputnik*, President Dwight D. Eisenhower, fearing that the Cold War would expand into space, established NASA (the National Aeronautics and Space Administration) in an effort to match — and outdo — Russian space accomplishments. Similarly, when Yuri Gagarin orbited the earth, President John F. Kennedy responded by promising to send a man to the moon, thereby creating a 'space race'.

What has this to do with the history of technology? Throughout most of history, scholars — in both East and West — had paid little heed to technological developments, except in their studies of the growth of states and empires through military might. However, the great advances in agriculture and industry during the eighteenth and nineteenth centuries heightened interest in the 'whys and wherefores' of technological change, as nations sought to better their economic condition through increases in production of both the necessities and pleasures of life. Furthermore, the great destructive capabilities of technology, manifested during two world wars, seemed to demonstrate that a nation's future existence, as well as its capabilities for economic improvement of the life of its people, depended upon its mastery of science and technology.

Thus historical scholarship in the 'groves of academe', which earlier had concentrated almost solely on political and military matters, began taking increased cognizance of scientific and technological developments. But inasmuch as great advances in electrical and chemical sciences during the nineteenth century had helped bring about major technological changes, there was a tendency to view technology as merely 'applied science'. Hence the history of science tended to dominate studies of the technological past — at least in the United States, where George Sarton, creator of the History of Science Society and its journal *ISIS*, established an academic programme in the history of science at Harvard. So the first generation of scientific historians in the United States concentrated on scientific developments alone, while tending to view the history of technology with disdain — or, as they quoted Sarton, limited their concerns to 'thinkers, not tinkers'.

Nevertheless, the introduction of the history of science to academia indicated that the simplistic view of history as 'past politics', which had prevailed throughout the nineteenth century, no longer sufficed. Instead, it was recognized that many different factors — including economic performance and sociocultural ideologies — strongly influenced political structure and behaviour. And among the most significant of these factors were the technological developments which brought about industrialization — and hence changed where and how people lived, worked, thought, played and prayed!

In the United States, just as technical education had been expanded in the nineteenth century to include more than 'shop' training by the addition of mathematics and sciences — a reflection of the growing 'scientification' of engineering practice and devices — so during the twentieth century there were discussions of the need to broaden technological education beyond the 'nuts and bolts' of technical devices by giving engineering students greater acquaintance with the liberal arts, especially the humanities and social sciences. Thus, following World War II, the Engineers' Council for Professional Development, at that time the accrediting agency for engineering educational institutions in the United States, recommended that studies in the liberal arts should constitute as much as 12.5 per cent of an engineering curriculum — and the ASEE (American Society for Engineering Education), the professional organization of engineering educators, created a Humanistic–Social Division (H–S Div.) to see how those new liberal arts requirements might best be met.

In the 1950s, when I became involved in teaching the humanities and social sciences to engineering students, the term 'general education' was the latest 'catchword' in American institutions of higher learning — for Harvard University used it to describe the breadth of its liberal arts studies, instead of the narrow specialization of individual disciplines. One of the best ways to achieve 'general education', it was claimed, was to introduce courses on 'Western civilization' which would weave together political, artistic, literary, philosophical, economic, scientific and social developments into an 'integrated cultural pattern'. Then, utilizing Matthew Arnold's nineteenth-century definition of 'culture' as 'the best that man has thought, said and done' in the past, we historians — with the aid of our colleagues from related humanistic–social science disciplines — put together some rather impressive courses.

While these innovative courses kindled the imagination and ingenuity of the faculties in the humanities and social sciences, our engineering students, alas, remained unimpressed by these new and innovative courses in the engineering curriculum. Why? My guess was that there were no figurative dollar-signs in front of the course numbers. What do I mean? Well, teaching at engineering institutions, where the students are very much 'career-oriented', I realized that while they rightly believed that their science and engineering courses would have a close bearing upon their professional careers, they wrongly believed that courses in the humanities and social sciences were irrelevant to their futures. How could we overcome that problem?

Then I had the good fortune to remember a statement by John Dewey, a great American educator during the first part of this century: 'To get a student interested in something, start him out with something he is already interested in!'

So, since our students at technological institutes were majoring in science and engineering, we changed the approach of our humanistic–social courses from Matthew Arnold's nineteenth-century view of 'culture' to an opening question which might provoke the interest of our students by relating history directly to their field of interest: 'How have science and technology influenced the development of Western civilization?'

Then, starting each part of the course from antiquity to the present, with that question, we would then move on from the scientific–technological developments to the political, military, social, economic and cultural aspects of the historical era and place them under consideration. Attempting to show the interrelatedness of scientific–technological developments to other aspects of society and culture, we became increasingly aware of the need to study scientific–technological developments in their sociocultural context. We, the faculty, learned *that* while teaching it to our students — and then we proceeded to apply that contextual approach to the history of technology itself.

At first, as indicated above, we hoped that the history of science, already a flourishing field in the United States, would broaden its agenda to include the history of technology. But when historians of science 'looked down their noses' upon the history of technology, we created a new organization, the Society for the History of Technology (SHOT) and a new journal, *Technology and Culture*, to foster study in that field – and also to enlarge comprehension of what the great philosopher, Immanuel Kant, some two centuries ago, had stated to be the purpose of historical learning, namely (in a very loose translation): (1) to find out how things got to be the way they are, and (2) what we might do about it!

In brief, Kant pointed out the relevance of studying the past in order to comprehend today's world – and, very importantly, help us cope with the problems of today and tomorrow. And the fact that historians of technology – at least in the United States – became increasingly involved in discussions on science/technology policy, sociotechnical parameters of environmental problems, socioeconomic questions posed by the introduction of new technologies, and the like, indicated that studying the history of technology is not the pursuit of recondite knowledge. Rather, it provides meaningful inputs to assist in public policy, societal activities and the entire range of human concerns, so as to produce a better world for tomorrow.

But I have got ahead of my story. Let me go back to the year 1958/59 when SHOT was officially founded and the first issue of *Technology and Culture* was being prepared for publication (January 1960). As editor of that journal and as secretary of the organization, I felt that I should have more knowledge of what was being done in other countries with regard to the history of technology. So I attended the IXth International Congress of the History of Science, in Barcelona and Madrid, for I knew that the International Union of the History of Science, unlike the History of Science Society in America, *did* include technology among its concerns.

At that time, the International Union was quite small. Only 108 papers were presented at the entire congress, *but* twenty-one of them dealt with aspects of *technological* history. Not only did I use the opportunity to acquaint some of my American colleagues with SHOT's forthcoming journal, *Technology and Culture*, but, perhaps more importantly, I met two distinguished, historians of technology from Eastern Europe: Eugeniusz Olszewski of Poland, who was to be the chief figure in the establishment of ICOHTEC, and A.A. Zvorikin, a pioneer historian of technology in the Soviet Union. Their friendliness made me realize that the warmth of scholarly companionship need not be affected by the political chill accompanying the Cold War.

In 1962, when the Xth International Congress of the History of Science was held at Cornell University (Ithaca, New York), an entire section of the programme dealt with the 'History of Technology and Applied Science'. Some twenty-six papers were presented, ranging from antiquity to the development of the aircraft industry in the twentieth century. Quite a few Americans participated, including

John B. Rae (one of my co-founders of SHOT), and there were also some distinguished foreign participants, such as A.G. Drachmann (Denmark), who spoke on 'How to Write a Paper on Ancient Technology', and Bertrand Gille (France) on 'Sources for the History of Technology in the Middle Ages'.

It might be said that the impetus for founding ICOHTEC derived from this meeting. For it enabled me and three major European historians of technology to come into closer contact: Olszewski, who presented a paper called 'The Scope of the History of Technology and Its Place among Other Sciences'; Semyon V. Shuchardin, the prime historian of technology in the Soviet Union, who gave a paper entitled 'On the Problem of Motive Forces in the Development of Technique'; and Maurice Daumas (Centre National des Arts et Métiers, Paris), whose 'La Mythe de la révolution technique' pointed out that mechanization had began much earlier than the Industrial Revolution, with, for example, textile-making innovations going far back in history.

At this Cornell meeting, the four of us spoke with one another about the desirability of an international organization to bring together historians of technology on a world-wide scale, just as the International Union was doing with the history of science. Following the Cornell meeting, we continued our discussion by mail — and that is what impelled Olszewski to take steps to organize such a group when the International Union held its XIth International Congress in Poland (Warsaw and Cracow) in 1965.

The programme of that 1965 Congress indicated that the history of technology was indeed becoming a major field of study. Thus a number of Congress sessions dealt with the 'History of Technology and Applied Science', starting with one on 'General Problems', chaired by Olszewski, and with papers by Shuchardin and Kranzberg. A second session, chaired by Kranzberg, dealt with 'Diverse Technologies (until the eighteenth century)' and had papers by such luminaries as Joseph Needham, whose pioneering studies in the history of Chinese science and technology had already brought him international fame; Sidney Edelstein (who later established the Edelstein Center on the History and Philosophy of Science and Technology at the Hebrew University, and who funds SHOT's Dexter Prize for the best book in the history of technology) spoke on the dyeing and tanning of leather in the sixteenth century; and Ladislao Reti, whose paper on 'The Codex of Juanelo Turriano (1500–85) and its importance for the History of Technology', also had indirect importance for the history of technology by attracting José-Antonio García-Diego, who later became ICOHTEC's president and benefactor.

All told, some fifty-five papers at the Warsaw Congress dealt with aspects of the history of technology, thereby providing a strong argument for establishing a separate section on the History of Technology within the International Union. The actual steps for its

Eugeniusz Olszewski

Maurice Daumas

Semyon Shuchardin

Melvin Kranzberg

creation were taken when Olszewski invited Daumas, Shuchardin and Kranzberg for dinner at his home. It was a sumptuous dinner, and it has stayed in my mind for several reasons. For one thing, the main course was roast goose, with all the trimmings, prepared by his wife, Barbara – and while I had often eaten roast duck, I had eaten roast goose only a few times – and it was absolutely delicious. Second, it was in Olszewski's home – and while I already had some close friends in Eastern Europe, this was the first time I had ever been invited to one of their homes. (I realize that social customs in Europe are somewhat more formal than in the American Midwest, where I was born, and the American South, where I have lived for two decades.)

At this dinner meeting, we decided to move ahead with formation of an organization, ICOHTEC, and to achieve its recognition as a Scientific Section of the International Union in time for the next Congress, scheduled for Prague in 1968. We also decided on an organizational form which, while tacitly recognizing the national fears and suspicions aroused by the Cold War, would ensure that ICOHTEC would not become involved in the political differences among nations and hence would not allow those rivalries to affect our scholarly mission. After all, we had met one another at previous IUHPS congresses, and since the United Nations, in general, and UNESCO, in particular, provided respectable mechanisms which could function across national boundaries, especially at a time when the Cold War was going on, affiliation with IUHPS was the obvious path to take.

In that connection it should be pointed out that while independent scientific/technical scholarly organizations, quite separate from state institutions, are prevalent in the United States and (to a somewhat lesser extent) in Western European nations, such independence did not exist at all in Eastern Europe at that time. Thus, of the four historians at Olszewski's dinner meeting to set down organizational lines for ICOHTEC, both Schuchardin and Olszewski were attached to the historical sections of their countries' national academies, while Daumas was curator of the Conservatoire Nationale des Arts et Métiers in Paris, which is under governmental control and is the premier centre for the history of technology in France. At the time, I was teaching at Case Institute of Technology (Cleveland, Ohio), a completely private institution which had no direct funding by any governmental institution. My role of vice-president of ICOHTEC derived from my SHOT affiliation, for SHOT represented America in the ICOHTEC General Assembly, which at its outset was organized so that each nation had one vote – and without individual members having votes in it.

Besides, at this time of heightened Cold War tensions, we felt that the only way we could enlist international collaboration for an institutional mechanism to promote the history of technology was through the United Nations – in particular, UNESCO (of which IUHPS is a part). And, as far as ICOHTEC's own organization was concerned, we wanted to make certain that it would not be a 'battleground' between the great powers, so we arrived at a tacit agreement that the ICOHTEC president would *not* be from the USA *nor* the Soviet Union, but could be from any other country. However, there would be two vice-presidents, one each from the United States and the Soviet Union. And, in order not to exacerbate sensitivies already heightened by the Cold War, the two vice-presidents would be equals, with no hierarchical distinctions in titles, such as 'First Vice-President' and 'Second Vice-President'.

Another safeguard to prevent ICOHTEC from becoming a 'battleground' in the Cold War was embodied in the ICOHTEC Constitution, which did not give ICOHTEC's president the power to 'run' the organization; instead, that was the province of ICOHTEC's General Assembly, which met every three years when the International Congress was held. Several years later in 1977 the International Union changed its congresses to a four-year interval between meetings.)

Another tacit agreement about ICOHTEC's officerial organization derived from the difficulties involved in currency exchange with Eastern European countries at that time. So it was decided that ICOHTEC's secretary general, who also serves as its treasurer, must be from a country where there are no barriers to international money exchange. And inasmuch as the secretary-general was also entrusted with producing a newsletter from time to time, he must be from a country where correspondence abroad, including newsletters, would not undergo government censorship. When all those factors were put together, it meant that the secretary-general would have to be from a Western nation.

The above rationale for selection of ICOHTEC officers seems to have been tailor-made to fit those at Olszewski's home for dinner in Warsaw in 1965. Olszewski, who had set in motion the formation of ICOHTEC, rightly became its first president. Shuchardin and Kranzberg became vice-presidents, and Daumas became secretary-general. Leaving personalities aside, one can see that this organizational blueprint worked very well, for it meant that the Cold War was never incorporated into the structure of the international community of historians of technology, and it protected ICOHTEC's funds and the secretary-general's freedom of expression.

(Coincidentally, ICOHTEC's vice-presidential arrangement imitated the structure of the United Nations, where below the UN Secretary-General are two top-ranking officials: one, the Deputy Secretary-General for General Assembly Affairs always being an American, while the Deputy Secretary-General for Security Council Affairs was always a Soviet. Theoretically, both deputies are independent of their governments, but that has not always proved so in practice. In the case of ICOHTEC, however, to the best of my knowledge, neither Russian nor American government agencies sought to utilize their respective ICOHTEC vice-presidents for pursuit of political ends. They probably thought that we were engaged in activities that were not worth worrying about.)

When the International Union accepted our petition to make ICOHTEC a Scientific Section of IUHPS, ICOHTEC's first symposium was to be in conjunction with the XIIth International Congress of the History of Science in 1968, scheduled for Prague. But 1968 was also the year of the 'Prague Spring', when Alexander Dubček sought to free Czechoslovakia from Russian domination. Fearing violence and disruption in Prague, the International Union moved its Congress to Paris.

At this first ICOHTEC symposium in Paris, we worried that the troubles in Prague would prevent our Czechoslovakian colleagues from participating. Hence I shall never forget the wave of applause and standing ovation given Mikulas Teich (a native Czech, now at Cambridge University), who was able to escape Prague, when he showed up at our symposium in Paris. (Actually, not all our Eastern European friends joined the standing ovation; indeed, many of them looked somewhat embarrassed when Teich appeared.)

This first ICOHTEC symposium was extremely successful, for it showed the wide range of topics which formed part of the history of technology and enlisted both well-known scholars and some young ones who were later to become 'stars' in the field. For example, R. Angus Buchanan spoke on 'The Contribution of Industrial Archaeology to the History of Technology'. At that time, industrial archaeology was a relatively new field — and somewhat 'imprecise', as Buchanan indicated in his paper. But his paper — and his subsequent work in that field — helped give it precision and indicated its importance as a means of recording and conserving technological source-materials which enable historians to 'grapple with some very practical problems of power supply, fuels, mechanical construction, metallurgical details, chemical processes, and dynamics', thereby conveying knowledge of the many details involved in technological history and a heightened understanding of processes of technological change. (Buchanan's many contributions to scholarship in the field of industrial archaeology were recognized when Queen Elizabeth bestowed the Order of the British Empire on him in 1992, and his eminence in the history of technology was shown by his election as secretary-general/treasurer of ICOHTEC in 1981, and his elevation in 1993 to the ICOHTEC presidency.)

Chronologically, the topics at this first official ICOHTEC symposium in 1968 ranged from antiquity — for example, Henri Goblot (France) on the draining of mining galleries from ancient times in the Tigris-Euphrates valley, and Isa Rizajevitch Selmkhanov (USSR) on bronze production in the ancient Caucasus — through the Middle Ages, such as Marjorie Boyer (USA) on 'Bridges and Mill Sites in Medieval France' — then to early modern times, such as Jiri Majer (Czechoslovakia) on Agricola and mining technology in central Europe in the sixteenth Century, and Ladislao Reti (Argentina) on

Juanelo Turriano's horizontal waterwheels (*c.* 1565) — through the beginnings of industrialization in various parts of Europe, and showing how ideas and techniques were transferred from different parts of Europe to another.

The origins of some major components of modern technology were also dealt with, as, for example, Luigi Bulferetti's (Italy) paper called 'The Italian Correspondents of Charles Babbage'; Otto Mayr (USA at the time, but later head of the Deutsches Museum in Munich) on 'Feedback Devices and Liberalism in the 18th Century'; Charles Susskind (USA) on 'Relative Roles of Science and Technology in Early Radar'; and Barton C. Hacker (USA) on 'The Genesis of Project Gemini: The Idea of Orbital Rendezvous, 1929–61'. There were also many papers dealing with the transfer of technology from one nation to another and the transfer of technological knowledge, as for example, František Jelek (Czechoslovakia) on the relations of the École Polytechnique of Prague with its Paris counterpart.

The above papers — plus a whole host of others — covered many different fields of technology, a wide chronological range, as well as the geographic diversity of technological developments. These different approaches in time, place, field of technology, and the like provided a true *raison d'être* — a justification, a rationale — for the existence of an international organization, such as ICOHTEC, to foster study of the history of technology throughout the world and to bring together scholars from different countries to understand the international nature of technology, its derivation from many different places and at many different times, and its contributions to human beings on a global basis. And the fact that the history of technology was not just an arcane study of the past, but was extremely relevant to present and future developments, was also evident in the paper presented by one of the first Asian scholars to participate in ICOHTEC, Matsutomo Yuasa (Japan): 'The Role of Science and Technology in the Economic Development of Modern Japan'. In brief, this first ICOHTEC symposium showed that professional historians already recognized that they have much to learn from one another by actively exchanging knowledge with their colleagues on an international basis.

Not only did participation in ICOHTEC broaden our minds, but it also broadened our hearts. Let me explain.

When in 1970 ICOHTEC held its second symposium, the topic was the acquisition of various technologies by countries which had not initiated them. This was ICOHTEC's first self-standing symposium — i.e. not as part of a larger congress of the International Union — so it did not have to be in a large city with facilities for hundreds of participants. Instead, it was held in a former monastery, built in the Middle Ages (but much modernized since then) in Pont-à-Mousson, a pastoral village not far from some large factories which we also visited in the course of the symposium.

This ICOHTEC symposium not only established the practice of meeting and wining-and-dining in beautiful historical locations, but it also established a pattern for including sightseeing trips and special social events in the scheduled activities – in addition, of course, to the scholarly symposium sessions. Among the social activities at Pont-à-Mousson was a reception at the Mayor's Palace in nearby Metz, where we were greeted by the Mayor and treated to rounds of drinks and delightful food delicacies. This type of reception at a major public building or museum with an official 'welcome' by a mayor, governor, or other major public official was to characterize all future ICOHTEC meetings – and, of course, that accounts for the special nature of ICOHTEC gatherings, which far surpass those of most other organizations and which account for the warm friendliness which prevails among all the participants, no matter what their national origins.

What made ICOHTEC's visit to the mayoral palace at Metz especially meaningful to me was that during World War II, the US Army unit to which I was attached (347th Regiment, 87th Infantry Division, Third Army) was first committed to action there. The Third Army had besieged Metz in order to oust the German forces occupying the city, so that we could push on to the nearby Saar valley and thereby wrest a major mining and industrial centre from enemy control. Hence it was at Metz that I first came under enemy fire from the forces of the Third Reich.

My ICOHTEC colleagues at the Mayor's reception included scholars from both East and West Germany (the German Democratic Republic and the Federal Republic of Germany). As I looked about me I realized that, in the autumn of 1944 when I was exchanging gunfire with German soldiers in Metz, I never would have thought that a quarter of a century later I would be at the Mayor's Palace in Metz with some German colleagues – and we would happily be drinking toasts to one another!

It was a most touching experience for me. Not only did it remind me of the brotherhood of all mankind, but it was also in accord with the biblical injunction against holding the sons responsible for the sins of their fathers. While I can never completely forget some of the pains visited upon me during combat, ICOHTEC effectively ended World War II for me – and it also meant that I never allowed the Cold War to take hold of me, at least in the form of affecting personal friendships.

I am quite sure that I was not alone in that experience. For one of the finest things about ICOHTEC is that it brought together in friendly fashion individuals whose governments were not always kindly disposed towards one another. Thus Germans from both East and West fraternized with one another at ICOHTEC – and Czechs, Poles, Russians, Frenchmen, and citizens of other nations who had participated in World War II felt no animosity toward their scholarly colleagues from Germany, nor did the Germans display any feelings of resentment. In brief, ICOHTEC has placed all of us together — on the 'winning side'.

However, almost two decades later, at ICOHTEC's XVth Symposium in Dresden, in 1986, I did receive a somewhat rude and jolting reminder of World War II when, in the course of the wonderful cultural activities which always accompany ICOHTEC meetings, I saw certain plaques on the recently reconstructed opera house and on the former palace which had become a beautiful museum housing many art treasures. The plaques (in German) read as follows: 'Destroyed by American and British aerial bombardment; reconstructed with the aid of our Russian comrades.'

It is certainly true that Dresden underwent very destructive bombing by the US Army Air Corps and, to a lesser extent, from the RAF, near the end of World War II. At that time my Third Army unit was rushing across southern Saxony to the Czech border — and it was my impression that the Russians who were fighting their way towards Dresden were fearful that they would be met with tremendous last-minute resistance as the Wehrmacht sought to defend one of the last great bastions of German might and power. Hence I was under the impression that the Allied air forces, having already destroyed Germany's major industrial centres and main transportation lines, had bombed Dresden on behalf of — and at the request of — our Russian allies, in order to protect them against further casualties in a bloody combat which had already caused them millions of casualties. (Later on, following the war, it was revealed that Dresden had become a major place of refuge for Germans who had fled from their homes in areas which were being occupied by American or Russian forces — and that Dresden did not possess the human or military resources to undertake a strong defence against Allied ground forces. Hence the American and British bombing of Dresden had visited great destruction on defenceless civilians and served no military purpose. Since I was under the impression that the bombing raids had been undertaken at the request of the Russians, I found the plaques on the opera house and the palace to be somewhat ironic — as well as a measure of the extent to which the East Germans joined with their Russian 'comrades' in the Cold War against the West.)

Nevertheless, even when the Cold War between East and West began, the national animosities were *not* transferred to the individual scholars comprising ICOHTEC. Yes, there were occasional reminders of it, but these were minor. For example, the Russians gave me several enamelled pins commemorating Yuri Gagarin's pioneering space-circumnavigation of the Earth. I thanked them — but I never wore such pins anyhow. When in 1969 Neil Armstrong landed on the Moon, I brought panes of United States postage stamps depicting the Moon landing to the next ICOHTEC meeting and gave them as gifts to my Russian friends. (Of course, they thanked me profusely, just as I had earlier thanked them for the Gagarin pins!)

But the fact is that, in addition to such superficial expressions of friendliness, ICOHTEC truly developed a feeling of scholarly comradeship. Partly this was owing to the fact that ICOHTEC meetings, in addition to the scholarly sessions, always included a number of social gatherings, with wine and beer flowing freely. As a result, ICOHTEC meetings became known for the *gemütlichkeit* (good-natured, kindly friendliness) which prevailed. And since ICOHTEC held its own symposia in the years between the International Congresses, as well as at the Congresses, this meant that ICOHTEC members saw one another more frequently than did the members of most other international organizations — and that, too, contributed to the friendliness.

Were there never any exceptions to the warmth and friendliness described above? Didn't the Cold War ever enter into the 'business' of ICOHTEC itself or in the scholarly papers at its meetings? Surprisingly, given the temper of the times, my answer to those questions is, 'For the most part, no' — but I do recall a couple of exceptions.

Thus when ICOHTEC went through the agenda at its business sessions, one occasionally might have felt that political affiliations mattered. For example, when the question came up about deciding which country's invitation to accept for a forthcoming ICOHTEC symposium, I was left with the impression that the delegates from the Eastern European bloc of nations waited until the delegate from the Soviet Union expressed his opinion. Once he did, representatives from East Germany, Poland, Czechoslovakia and the other Eastern European nations would then get up and echo his remarks.

But did ideological differences ever impinge upon scholarly discussions? While sociopolitical ideologies might certainly have affected the choice of scholarly topics and the approaches to them, frankly, I can recall only one time when this was patently the case — and that might be because it happened to me. It was when I presented a paper (at an ICOHTEC Symposium in East Germany) which dealt with improvements in working conditions in factories made possible by the rapid industrialization in the nineteenth century.

My point was that the great growth of productivity, made possible by technological advances, meant that the number of working hours per week could be cut, and that the use of women and child labour, especially in dangerous places such as coal-mines, could be abolished. And I cited production facts and figures — and also British parliamentary debates — in the nineteenth century which led to the abolition of women and child labour in mining and other dangerous industries. I also pointed out specific legislation, such as the 1833 Factory Act in Britain and later legislation in Britain along the same lines, as well as Bismarck's (1883) 'state socialism' measures in the German Empire to insure workers against sickness and accidents.

Much to my surprise, at the conclusion of my paper, several East German scholars got up and began to dispute my paper, arguing that the capitalists who controlled the British Parliament had refused to do anything on behalf of the workers' safety, health and comfort. And they cited statistics which 'proved' their points. When I asked them whence they had derived their information, they told me that an abundance of facts and figurers could be found in Friedrich Engels's study, *The Condition of the Working Class in England in 1844*. That study, of course, was written before many measures of social legislation enacted by the British Parliament had begun to take effect.

All I could do was shake my head in disbelief, and point out that much of the legislation on behalf of the workers undertaken by the British, German and other European governments was aimed at undermining the appeal to the working classes of the socialist writings of Engels and Marx. And those social measures had worked, for communist revolutions did *not* occur in the industrialized Western European nations where Marx and Engels expected the communists to take over — but, rather, in industrially backward, agricultural regions such as Russia, where Marx and Engels thought communism, as they viewed it, could *never* take hold. (As might be expected, my critics shook their heads in disbelief at my defence.)

However, such disputes founded on sociopolitical ideologies were very rare in ICOHTEC. The scholarly papers were primarily based on actual historical data of a non-sensitive kind, so we averted political discussions of the above nature.

Furthermore, the basic structure of ICOHTEC, expressed in its written constitution and in the tacit agreements regarding the countries of origin of its officers, remained in effect without any questions being raised. For example, when Maurice Daumas stepped down as ICOHTEC secretary-general in 1974, his place was taken by John Abrams, Director of the Institute for the History and Philosophy of Science and Technology at the University of Toronto. Although born and educated in the United States, Abrams was so intense in his opposition to Nazism that he moved to Canada at the outset of World War II — some two years before the United States became involved — in order to enlist in the Royal Canadian Air Force. Although possessing degrees and background in astronomy and engineering, Abrams was deeply interested in the interactions of science, technology and society throughout history, and his post at Toronto made him well suited to be ICOHTEC secretary-general. And, in accordance with ICOHTEC policy of keeping its funds in Western banks, he put the money in Canadian banks. Upon his death in 1981, R. Angus Buchanan of the University of Bath succeeded him, and the money was transferred to banks in West Germany and England.

When in 1974, Olszewski stepped down from the ICOHTEC presidency, after serving six years, he was replaced by Luigi Bulferetti (Italy), 1974–7, with Carlo Macagni (Italy) replacing him (1977–81). This was a time when interest in technological history was growing in Italy, and, indeed, the Italian Council of National Research had funded the 1973 ICOHTEC Symposium on the 'Industrialization of the Agricultural Countries of Southern and Central Europe during the 19th Century', held in Jablonna, Poland in 1973. Again, it was a very successful meeting, held in a lovely eighteenth-century château not far from Warsaw. Unfortunately, the Italian Council had *not* sent the subsidy for publication of the proceedings to the ICOHTEC secretary-general, but directly to a Polish bank — and the thousands of zlotys deposited there could not be transferred to a Western bank. So the Jablonna proceedings were never published, for, by the time they were ready, the zloty had fallen so far in value that there was not enough money in the account for that purpose. But the Jablonna meeting was very successful in terms of scholarly learning and goodfellowship. And the social occasions included an evening visit to Chopin's summer home, where a fine musician played a selection of beautiful études on Chopin's own piano.

ICOHTEC continued to flourish. Indeed, at its fifth symposium, held in Tokyo (1974) in conjunction with the XIVth International Congress of the History of Science, there was a noticeable broadening of the geographic scope of the history of technology to include Japan and other Asian nations. This included the early contributions of Asiatic nations to technology and the later transfer of Western industrial technology to the Far East. While only a few Asian scholars had previously participated in conferences on the history of science and technology, their number has grown since then, thereby expanding ICOHTEC's membership as well as its geographic scope. There were also papers on technology's environmental impacts, a topic which has increasingly engaged scholars since then.

Two years after that historic Tokyo Congress, ICOHTEC held one of its most distinctive meetings in Kaluga (USSR). At least I regarded it as distinctive for several reasons. For one thing, Kaluga is far from the beaten path of usual tourist travel in Russia. Indeed, the official Russian travel agency for foreign visitors would not make arrangements for people to travel there. Instead, the Russian Academy of Sciences had a bus take all sixty participants (representing eight countries) from Moscow to Kaluga, where we stayed in a small but beautiful hotel with better accommodations and meals than the fancy hotels in Moscow and Leningrad.

Why was Kaluga chosen for the site of this ICOHTEC meeting? I suspect that it was because it had some worthwhile places to visit that are not normally on foreign tours, and also because it provided evidence of Russian contributions to scientific and technological advance of which most Western scholars are unaware.

Both the above were embodied in our Kaluga visit to the home of Konstantin Tsiolkovsky, a Russian scientist. Although not many people outside of Russia had ever heard of him, I happened to know a little bit about the history — and, indeed, the prehistory — of human space flight, for at that time I chaired the History Advisory Committee of the National Aeronautics and Space Administration (NASA). At the beginning of this century (1903), Tsiolkovsky provided mathematical proof that it was possible to use the reactive force of rockets to eject a vehicle into space beyond the pull of Earth's gravity. Tsiolkovsky was correct, but even though the Romanian Hermann Oberth in the 1920s made similar mathematical studies, and the American Robert Goddard (in the 1920s and 1930s) strove to produce a rocket that would permit travel to outer space, it took over half a century before *Sputnik* actually circumnavigated the globe in space.

Along with the visit to Tsiolkovsky's home, the Kaluga trip included a sightseeing visit to a nearby space museum that indicated Russia's early leadership in thinking about — and performance in — cosmonautics. As an American I might have claimed that we had 'defeated' the Russians in the 'space race', but I was truly impressed by the visit to this Russian cosmonautic museum for it showed the contributions made by many different peoples which enabled Neil Armstrong to take 'one giant leap for mankind' when Apollo 11 landed on the Moon and he 'Moon-walked' in 1969.

In 1977, the year after Kaluga, ICOHTEC had two symposia in the United Kingdom. One, in conjunction with the XVth International Congress of the History of Science in Edinburgh, had papers dealing with various aspects of technological history, especially on the contextual interactions of technology with science, the sciences of man and philosophy. The second symposium, immediately after Edinburgh, was in Stirling, in conjunction with a Newcomen Society meeting on Scottish engineers and engineering, and it gave ICOHTEC members the opportunity to focus on the 'internal' history of technology.

ICOHTEC's tenth anniversary (1978) meeting was sponsored by Bergakademie Freiberg, the world's oldest mining academy, and some eighty participants presented scholarly accounts on various aspects of the history of mining and of research on various materials. The programme sessions were held at the Bergakademie, but the visitors stayed at a beautiful castle dating back to medieval times — and had medieval feasts (including a 'cook-out' on the filled-in moat of the castle) and medieval tournaments, including an archery contest!

Because I knew that the Bergakademie had the oldest mining museum in existence, with collections of materials excavated from mines throughout the world, I was prepared to be impressed by the collection — and indeed I was. But I had also brought with me a 'surprise' to add to the collection — and I presented it to the head of the Bergakademie at the final banquet. What was it? As I indicated above, I was deeply involved with America's space agency, NASA; and in order to reward me for my work as chairman of its History Advisory Committee, NASA presented me with an Apollo Achievement Award (a certificate which hangs in my hallway at home) and a completely unexpected 'gift from outer space': a medallion, celebrating Apollo 11's 'Eagle' landing on the moon and containing a small sample of the Moon's surface. It might have been only a molecule or two, I don't know, but I do know that the Apollo astronauts had brought back samples of the moon's surface, which were being subjected to close scientific scrutiny — so I assume that some 'leftover' bits and pieces, which were not needed for further scrutiny, were melted together with metals from the Earth to produce medallions containing a grain or two from the moon.

I was sent two of those medallions, and I still retain one. But what better use could I make of the other than to present it to the world's oldest mining museum at the world's oldest mining academy? There it would serve as a symbol of ICOHTEC's appreciation of the Bergakademie Freiberg's hospitality on our tenth anniversary, the large role played in ICOHTEC by our East German colleagues (e.g. Eberhard Wächtler of the Bergakademie faculty), and ICOHTEC's role in bringing scholars together in the common pursuit of knowledge and in friendly *gemütlichkeit* despite the political challenges raised by the Cold War. I think that everybody was impressed by my 'heavenly' addition to this great collection of mining materials extracted from all portions of our planet Earth.

Alas, after my triumphant, 'outer space' addition to our Freiberg meeting, I failed to 'return to earth' for ICOHTEC's tenth symposium (1980) in Sofia, Bulgaria. Why not? Because Bulgaria had no diplomatic relations with the United States, so I could not obtain a visa for Bulgaria from its non-existent embassy or non-existent consulate in Washington. When I informed John Abrams, ICOHTEC secretary, about this he tried to see whether I could obtain a visa from the Bulgarian diplomatic office in Canada. Well, no relations with the United States meant no relations with American citizens either. So, while I had friendly relations at previous symposia with Bulgaria's representatives to ICOHTEC, Alexandre Valtchev and Dimo Demirov, I still could not obtain a visa. So the Sofia symposium was the only ICOHTEC meeting that I never attended.

When ICOHTEC met in Bucharest (1981) for its eleventh symposium, on 'Technology, Humanism, and Peace', in conjunction with the XVIth International Congress of the International Union of the History of Science, Dr Stefan Balan (Romania) was elected ICOHTEC president. Balan served two terms, until 1989 (the International Congress had changed from meeting every three years to meeting every four years), and he was replaced by José-Antonio García-Diego (Spain). At that same Bucharest meeting, vice-president S.V. Shuchardin (Soviet Union), who had died, was replaced by Yuri S. Voronkov (Soviet Union), head of the Division of the History of Technology of the Soviet Academy of Sciences.

The contributions of García-Diego (ICOHTEC president, 1989–93) to the history of technology in general, to ICOHTEC in particular, and especially to individual members of ICOHTEC deserve special attention. For one thing, he was not an academician, museum curator, or government employee whose professional career involved the history of technology. Instead, his professional background was in civil engineering, where he was remarkably successful. But since his childhood he had possessed great curiosity about technical history, and it became a consuming passion when, in 1969, the Spanish Ministry of Public Works asked him to supervise the partial reconstruction of a landmark artefact in Toledo built by a great sixteenth-century maker of clocks and other devices: Juanelo Turriano of Cremona (Lombardy). The device that Turriano built in Spain, which first attracted García-Diego's attention to him, was a mechanism (approriately called the 'Artificio') which conveyed water from the Tagus river up to the Alcázar, the residence of the Spanish ruler.

García-Diego became so interested in the history of hydraulic works that by 1977 it had become his principal activity, and from 1985 it became his all-consuming activity. He concentrated at first on Turriano and also on Spanish scientists and technologists of the seventeenth and eighteenth centuries. His first involvement with ICOHTEC was at the 1971 symposium in Moscow, and he became a regular attendant and active participant in ICOHTEC. In addition to publishing a major study of Turriano, as well as a series of scholarly works on others who contributed to technical growth in the Enlightenment, García-Diego established the Fundación Juanelo Turriano, headquartered in Madrid, which publishes works on the history of technology, supports scholarship in that field, and, as we will see below, has assisted individual historians of technology in times of need.

In 1988 García-Diego organized the sixteenth ICOHTEC symposium in Madrid, on 'Civil Engineering between 1750 and 1850', along with a visit to Barcelona. And just as previous ICOHTEC meetings had involved special social occasions, García-Diego arranged for a reception in the garden of the Mayor's palace in Madrid, a special party in the Prado (!), and, in Barcelona, another special reception — and speech by the Governor.

While those represent García-Diego's fine attention to the 'business' aspects of ICOHTEC, as expected of all the organization's officers, he has made very special — and largely unpublicized — contributions on behalf of individual ICOHTEC members which have been of importance to ICOHTEC as a whole. Let me explain.

When a couple of years ago, the Berlin Wall came tumbling down — literally as well as figuratively — many valued ICOHTEC members from Eastern Europe began losing their positions as new administrators took over or, even if they retained their jobs, there were no longer funds enabling them to attend ICOHTEC meetings which entailed travel outside their country. Here the Fundación Juanelo Turriano came to their 'rescue', supplying them with funds which enabled them to attend ICOHTEC symposia in Vienna, Uppsala and Zaragoza.

Since the Fundación Juanelo Turriano is supported entirely by funds from only one donor, Señor José-Antonio García-Diego, it is he who enabled ICOHTEC to maintain its international character at a time when political changes, accompanied by financial problems, made it difficult for scholars from some Eastern European countries to participate in meetings outside their native lands. Alas, his death early in 1994 forced the Fundación to make some budgetary changes which limited its support of travel to ICOHTEC's 1994 meeting in Bath.

Fortunately, some of the financial burden of bringing scholars from Eastern Europe to the Bath meeting was assumed by Dr Sidney M. Edelstein, founder of the Dexter Chemical Corporation (New York), whose interest in the history of dyeing, alchemy and chemistry, from antiquity to the present, led him to accumulate a valuable collection of manuscripts and books. These are housed in the library at the Hebrew University of Jerusalem, which is also the home of the Edelstein Center for the History and Philosophy of Science, Technology and Medicine – which offers international fellowships for study in those fields (see page 144 above). Dr. Edelstein also initiated the Dexter Award in the History of Chemistry, administered by the American Chemical Society, and the Dexter Prize of the Society for the History of Technology, which is awarded annually for the best book published in the history of technology. Unfortunately, Dr Edelstein passed away in mid-September 1994, but it is hoped that the Dexter Corporation and the Edelstein Center will continue to support participation by Eastern European scholars in ICOHTEC'S good work.

Along the same lines, other members of ICOHTEC, including myself, have sought to assist our colleagues in the history of technology from the Eastern European nations to retain their positions in universities or in governmental academies when financial considerations forced the administrators to drop them from their institutions. For example, the costs to West Germany when it merged with East Germany meant that some prominent historians of technology were notified that they would lose their faculty positions at the end of the next academic year. With the assistance of my American colleagues in the history of technology, I instituted writing campaigns to help them retain their jobs. Thus I — and others — wrote to the Cultural Minister of Saxony, which was formerly within East Germany, in an effort to save the jobs — or the pensions! — of our ICOHTEC colleagues in those areas.

Similarly, when Czechoslovakia split asunder, our ICOHTEC friends in the Czech and Slovak Republics discovered that they would lose their posts in the scientific academies which had hitherto supported their work, and which were being forced to cut back because of financial restrictions. So we tried the same tactic with the heads of the academies of science as we did with the academic hierarchy, hoping that a display of the international significance of the work of their historians of technology would enable them to retain their positions.

In some cases, our efforts have worked, either directly or indirectly, and we have also endeavoured to obtain fellowships for our foreign colleagues from American foundations which provide funds for scholarly research in the United States. Alas, our efforts have not always been successful — but we *have* tried.

The growth of interest in the study of the history of technology and the enlargement of its scope were manifest at ICOHTEC's 1994 meeting in Bath. Not only were there 86 participants from 23 countries (including seven nations of the former Soviet European bloc), but at least a dozen women presented papers. Furthermore, these women scholars did not limit themselves to 'feminist' topics, as has been largely the case in the past; instead, they were involved in a wide range of technical history.

Part of this 'explosive' growth in the topics covered was owing to Brenda Buchanan's (UK) pioneering work on the history of gunpowder and her organization of an entire symposium section on 'The Manufacture and Marketing of Gunpowder'. As she pointed out in her introductory address on this topic, gunpowder's impact on technology was not limited to military usage, but also extended to civilian advances in mining, chemical, and other fields of engineering. Furthermore, inasmuch as different conditions governed gunpowder's production and use in different countries, the making and marketing of gunpowder had diverse economic, cultural, and social ramifications on an international scale.

Certain other parameters of technological history which were rarely addressed in previous scholarly meetings also became 'themes' at the Bath meeting. These included 'invisible technologies', some of which had been tried and found wanting, and others which were so much a part of our daily lives that we gave no thought to them – so that, as Carroll Pursell (USA) pointed out, the history of technology had not paid adequate attention to the 'experiencing' of technology, as well as the 'making' of it. And Graham Hollister-Short (UK) indicated that the incremental improvements in the crafts of an earlier age, which played so great a part in technical advance, were not usually recorded or discussed, so that this important element of technological history remained 'invisible'.

Not surprisingly, these and other views expressed at the Bath symposium gave rise to stimulating discussions, not only at the program sessions themselves, but also at the social occasions, including the 'table talk' at the luncheons and sumptuous banquets which form part of every ICOHTEC meeting. As a result, historians at ICOHTEC symposia learn much from their conversations at social gatherings as well as from the official programme sessions – and they relish the opportunity to exchange ideas with their colleagues from all over the world in such congenial settings.

The important thing to be learned from the above recital of the collegial relationships and personal friendships fostered by participation in ICOHTEC is that, just as technology itself is a very human activity, so is the history of technology. Hence the field of the history of technology, our organization, and the individuals comprising it are also subjected to the problems and worries affecting other organizations and human beings. But the warm friendships fostered by participation in ICOHTEC since its creation appear to be closer and go far beyond those of other international organizations.

The fact is that pursuit of the history of technology is enjoyable, and that historians of technology learn from one another – and have fun while doing so. As a result, warm feelings of personal friendship emerge.

Let me mention a personal anecdote, which would have been considered impossible when ICOHTEC was founded a quarter of a century ago. Dr Yuri Voronkov (my Soviet co-vice-president of ICOHTEC until my retirement from that position in 1989) has a daughter, Nastasya, who received a fellowship in 1992 to spend her junior year at an American university. That kind of undergraduate student fellowship in the United States for a student from the Soviet Union did not exist when ICOHTEC was founded – and it would have been impossible during the Cold War. But, of course, times have changed – so American students are welcome in Russia, and Russian students are welcome at American universities, with, of course, the major obstacle being the matter of funding.

In any event, Nastasya Voronkova, daughter of Yuri Voronkov, was awarded a fellowship for study in the United States during the academic year 1992–3. She chose to attend the University of Kansas. That is quite far from Russia, so she could not go home during the fairly brief Christmas vacation at American universities. So where did she spend her Christmas holiday? At the home of Mel and Les Kranzberg in Atlanta, George, USA! And we truly enjoyed her visit. For as Marshall McLuhan stated several decades ago, technology has made the world into a 'global village' in terms of transportation, communication, and human interdependency on a world-wide basis. And Nastasya's happy visit to the Kranzbergs shows that ICOHTEC has not only fostered the growth of scholarship in the history of technology, but has made historians of technology, no matter their national origins and religious or political affiliations, into one happy family!

Hamlet without the Prince: The Strange Death of Technical Skill in Histories

MICHAEL FORES

In the 20th century it has come to be regarded as almost axiomatic that technology is a rational activity stemming from the traditions of modern science.
Encyclopaedia Britannica on The History of Technology.[1]

* * *

He interrupted me.
'So you've swallowed all this, hook, line and sinker?'
'Not at all,' I said. 'I just wanted to know if there is any scientific basis for it.'
Corrigan snorted.
'Scientific basis, my foot!'
Agatha Christie, *The Pale Horse*.[2]

INTRODUCTION: MYSTERY ABOUT A BASE

Homo is an animal, and therefore an organism which has features common to organisms. It is sentient, being able also to react in certain ways to outside forces. It is something which changes and develops over time, something which matures, grows older and then dies. With the passing of time, as well, and like most of the other animals, *Homo* is capable of learning as an individual organism. Apart from all of these observable features, however — with the capacity to learn usually recognized by the teachers — if anyone of our species were to imagine the human characteristics of imagination and articulate speech onto one of the other animal species with the mixed fortune of sharing with *Homo* the single planet called Earth, this might well be what members of the other species would have judged to be the most characteristic of *Homo*: here has been the most forceful, wilful and meddling organism of them all. Here has been, for good or for ill, the species to have had the most influence on the single, named planet.

Homo has certainly not — in ways sometimes associated by members of our species with the thing(s) called 'science' and with a very oddly named 'technology' – achieved 'the conquest of nature': 'Technology has become [by 1968] a calculated and intentional application of scientific principles and data or "natural laws", in ways determined by smaller or larger groups of men . . .'.[3] Volcanoes still erupt, diseases still infect, and there remain what insurance companies call Acts of God. Yet a single organism on the single planet it calls Earth has observedly gone close, at times, to moving minor mountains there, in order to meet special and local requirements of examples of that organism. So this particular organism, *Homo*, whose members can be passionate and rational, serious and frivolous, callous and kindly, has put great ingenuity and effort into a range of enterprises whose participants have often chosen to rationalize as being 'humane', or 'pious', or 'progressive' or even 'rational' things to have done. Accordingly, and assuming the anthropomorphic award noted here, there is little doubt that the other species might have been keen to assume *Homo*, especially *Homo* acting in groups, to have had magical powers. That is to say, *Homo* must have seemed to them to have been acting forcefully within the single planet, in many and indescribable ways.

The affair (or perhaps the affairs) only commonly called 'technology' in English in the last few generations, has (have) provided some of the most dramatic examples of forceful *Homo* in operation actively. What might be called active 'Homocraft' has produced, or helped to produce, 'reclaimed' land from the sea, the making and use of many types of machine, means of habitation and transport. People who have called themselves 'engineers' for far longer than others have called them 'technologists', together with their own forebears in human specialism, have had much to do with contriving this and much more. *Homo*'s technical specialists, let us say, among whose number are those whom today's *logographoi* (writers of prose, as opposed to the poets[4]) want to make seem to have been producers of *logoi* like themselves, have managed to make, and get to operate effectively, a wide range of 'engines'. Citing a popular word, *Homo* has thus 'engineered' and contrived many changes in its milieu. And *Homo* has often apologized for such actions in ways listed above. Something called Industrial Management has been said to have been conducted 'rationally'. The goals of Economic Growth have been talked up to appear 'progressive'. The spreading of an allegedly Modern European culture has been done 'humanely'. And so on. Thus technical means have been put to use in support of a wide range of collective faiths from Christianity to Communism, from Development Economics to Capitalism. How has the oddly named 'technology' been treated in print for these sorts of affairs?

One theme about a shift in technical modes, being a theme echoed in other writings by English-speaking *logographoi* over the last generation or two, has been David Landes's, as first articulated in 1965 in the volume of the *Cambridge Economic History of Europe* given over to treating 'the industrial revolutions and after'.[5] Concerning

Technological Change and Development, in a chapter not only treating *logoi*, and later extended to produce *The Unbound Prometheus*, there was talk there of a type of 'marriage of science and technology' taking place during the nineteenth century in each of the leading countries. But this union was not an easy one to consummate: 'one really refers to a complex liaison, which was not consecrated at a moment in time but developed slowly and unevenly, and varies to this day from country to country . . .' The parties of this 'ever-closer marriage of science and technology' had experienced 'essential independence' during the time which the *Cambridge Economic History* has tagged as the Industrial Revolution; and, according to this thesis, even in union 'this is one marriage that requires permanent mediation to work; the gap between science and technology is far too wide for direct communication'.[6] So, to retain the original metaphor, the outsider to technical matters should consider four in a bed, with the other two participants named as Applied Science and Engineering, before any progeny can appear in a regime of what Cardwell has called that of 'technics based on science'.[7] But what identity should the reader assume for any of these four, human-like entities, given this: 'technology' sounds as if it were a part of science anyway (bi-ology, pharmac-ology); and some might have assumed that Engineering, Technology and Applied Science were barely distinguishable from each other (compare Forbes, quoted above concerning the conquest of nature, where 'technology' seems to have been an activity again, and an 'application' of science)?

In the particular exposition of a theme on paper, its author has stressed a continuing difficulty, in the conditions of what appear as an 'afterdom' (the aftermath of change) from the proposition itself, in effecting such things as 'ingenious applications of pure scientific principles to industrial needs'; and the main expressed focus has been claimed to lie with the actions of 'two activities'.[8] Yet the single English word 'science', as used by both the *logographoi* and others, has been most commonly tagged (in a metaphorical sense of a bearing across of meaning) to a body of knowledge (from the Latin root *scientia*, knowledge). Those who call themselves 'scientists' are normally very keen to tag things securely in aid of identification and future cognition. The first dictionary sense of 'science' is a body of knowledge, as is broadly confirmed by the special subject-matter of the 'discipline' named as the History of Science (the history of what *Homo* has claimed to know about things *Homo* has liked to call 'natural'). So the chances of anthropomorphism are strong here, and probably of the awe involved as well, when (say) an inert science-as-knowledge has been made, in print, to appear active like a human being. Futhermore, the *Cambridge Economic History*'s notion of technology-as-an-activity seems to run contrary to what may be assumed from phrasing such as this — 'We have the technology, but have we the will to use it?' — which tends to refer to hardware and systems to operate it. And, despite these problems of conception, expression and objectification (here, trying to represent an observed subject in words on paper), there is another one, which may be considered as a problem of distinguishing this *History*'s 'afterdom'.

The primary concern, regarding what has been given as a wedding with no known location and timing, seems to have lain with trying to track down any items which may have been observed to 'cross the gap . . . from technology toward science' and vice versa. By this precept, 'afterdom' should have been distinguishable, in technical terms, from 'the autonomous flow of scientific discovery [which] fed a widening stream of new techniques'.[9] And such a model of changing modes of 'technological change' (Landes's main expressed subject-matter), with an apparent accent on uses of science-as-knowledge, may have helped Musson and Robinson articulate this question about the technical conditions of the ill-named Industrial Revolution (a set of changes whose own 'afterdom' has been distinguished by a mainly urban service sector): 'Was it [British technological leadership], as Professor Landes has said, achieved through the effects of "practical tinkerers", or did it have some scientific basis?' (Elsewhere, these authors seem to have characterized a technical 'beforedom', by its *inter alia* 'practical, unregulated empiricism'.)[10]

Well perhaps so? Perhaps something(s) called 'science' has (have) a major part nowadays in altering technical activity irrevocably, through producing a 'base' in the form of articulated knowledge. But 'being scientific', for the forceful meddler, *Homo*, has involved providing some evidence from 'observation' (use of the human sense and aids to that use) to support even the most well-loved abstractions. And it is very clear, for those who have bothered to go out of their studies to observe active and skilful *Homo* at work, that *there is no way in which knowledge-use can be monitored*. For those who will be called engine-wrights here, there has never been any obligation to record every time such knowledge has been put to use — doing this would comprise meta-activity for engine-wrights, in the sense of the meta-world discussed below. And, anyway, increased working competence for active *Homo* typically results in less-overt knowledge use. Furthermore, if only because *Homo* has been a learning organism, and despite the claims made by latter-day *logographoi*, there has never been a discernible mode of action sensibly named as a 'scientific method', which might have provided a yardstick for 'scientific rationality' in human behaviour. Accordingly, there has been no way either to recognize expressly 'practical', 'empirical' and/or 'unscientific' behaviour, within the oddly named 'technology' or elsewhere. So the notions of a Scientific Technology and a Science-Based Technology can have no 'scientific basis' themselves, in the terms cited above from Agatha Christie, who would have understood such issues from her work as a pharmacist in World War I.

Citing her 'mystery' story again, the notion of 'afterdom' in technical activity discussed here must be set down as part of a 'pseudo-scientific, psychological set-up — full of modern jargon and reinforced by old superstitions'.[11] It is a pipe-dream of the *logographoi* who have reinvented technical activity, with *Homo* shorn of much of this animal's personal skill, as 'technology'.

In the companion volume to the *Cambridge Modern History*, the contributor on Religion and Secularization has had this to say: 'I shall consider secularisation primarily as the process of change from the interpretation of reality in essentially supernatural, other-worldly terms to its interpretation in terms which are essentially natural and focussed on this world.'[12] Science is an artefact, comprising *Homo*'s collected knowledge of that 'reality'. Though, to be sure, there are those who, by discussing 'scientific discovery' — 'Newton discovered the laws of motion' — have flirted with a confusion between object and subject, and have made the artificial seem to have been natural; and there has been the piety about the thing(s) called 'science' touched on above. Applied Science, Science-Based Technology, Industrial Revolution and Scientific Method are certainly not 'terms which are essentially natural' in the sense of the quotation from the *History*. The present paper will examine the proposition that much of the latter-day, generalized treatment of the 'relationship between science and technology',[13] two ill-defined entities, has only related to events in a meta-world separate from the single world of post-Newtonian conception. This has been a meta-world, in which *Homo* has appeared latterly as an animal strangely shorn of personal skill, with all the accent on knowledge processing, and for which the *logographoi*, knowledge processors themselves, have appeared as the gate-keepers, minding especially the sacred abstractions on which their beloved meta-world depends.[14]

In the following section, an attempt will be made to identify and name some of the key characteristics of engines and science, both artefacts. While engines are bulky and valuable things, made primarily for their utility at a cost, science weighs nothing, has no market value, and has been fashioned to be as close as possible to a notional truth about the single world and the universe about us. Though claims have often been made in recent generations about the special powers of an ill-defined 'science' — 'Science builds a new world!' — prudent people should neglect such pieties, which seem to have awarded life to something inert. They should look closely, instead, at products of human activity, which indicate differences in real-life working by scientists and engineers, as well as the unlikelihood of a 'science-based technology'. A model of active 'Homocraft' should help inform on these matters. By contrasting machine activity as, in the main, predictable force transmission, human activity can be seen to have specialized on risk-taking and dealing with what has been relatively unknown; so goals of 'rational', 'methodical' and norm-influenced conduct cannot be suitably ascribed to active *Homo*.

Science has only been made and maintained in ways that English speakers might call 'professionally' since the nineteenth century. The section '1874, Divergence and All That' (pp. 177–82) will deal with aspects of the attempts, made by the *logographoi* of the last few generations, to 'sell' the valueless science through the invention and propagation of the meta-reality already indicated. That has included the assertion of the conditions of a technical 'afterdom' also indicated, as often logged down using the metaphors of the engineers (structure, base, force, cause, designed to make sense of inert things in the main). An examination of real-life separation in the nineteenth century, and (say) the faith of 1874, to mark the founding of the separate Cavendish Laboratory in Cambridge, provides a better understanding of real-world affairs than a concentration on the opening in 1876 of Edison's Inventions Factory in the USA. The generalizations of the *logographoi* may have been more dramatic and theatrical,[15] more entertaining to read in their pseudo-scientific, jargon-ridden way (Christie, above), but they have not treated the past adequately.

SCIENCE AND HORSES AND ENGINES AND THINGS

Human beings have typically invented words, and put them to effective use as the names of things, in order to indicate thereby the identity of things, or of groups of things. Thus, 'tagging' has enabled human beings to specify and to particularize about things in their reach of comprehension. In both common and scholarly discourse, as conducted between a named and identified *Homo* M and *Homo* N (the two of them named separately, in order to indicate separate members of a teeming species), those humans involved will normally agree about the main subjects of concern, when statements such as these have been offered: 'X is black and costs ten marks'; 'Y belongs to class Z of things'; 'W is my next-door neighbour, to the left of my house looking from the street'.

In such a general fashion, words might be said to have been tagged to things (once more in a metaphorical sense, of a bearing across of meaning), or to groups of things, in aid of meaning as well as identification. Words are used thus in support of definition, through humans seeking to know the observed characteristics of named things, and so too seeking to tie them down as part of a process sometimes called profanation. Things said to be sacred in any particular culture are sometimes things whose names should not be uttered by pious folk; and they are often things which should not be awarded the indignity of close definition of their characteristics in the dis-secting (cutting apart) manner treated here. That is a manner in which natural things (gold, an oak tree) may be named and specified, as well as abstractions made by human agency (Statement A of science, laws). Similarly, Nordic Prince, a racehorse, from a breed of animal developed by *Homo*, is sensibly named in the general manner discussed here; and the naming is important to help prevent racecourse fraud with the illegal substitution of published runners (horses whose particular names have been published to appear in named races). The title of the present piece of writing, 'Hamlet without the Prince', used as a name in conjunction with its given subtitle, 'The Strange Death of Technical Skill in Histories', may serve adequately the purposes of identification and particularization implied from their use, as well as providing an indication of the piece's theme. The present Hamlet is, at least, the published runner here.

With this particular use of names, as indicators and informers about particular things (Nordic Prince, Statement A of science) or groups of things (horses, science conceived as a group of articulated statements), it is suitable to expand here on what has already been articulated about science. That is, science, as a made-up body of knowledge about aspects of the single world around *Homo*, about aspects of what the contributor to the *Cambridge Modern History* cited above has chosen to call 'reality'. That is, too, about science, as contrasted with engines and horses, different things, as indicated by a human use of different names for them.

With such a scheme in place for the use of names, *Homo* will have used what might be called the scalpel of dissection, with his or her conviction that Subject P and Subject Q are things separable enough from each other, to deserve the award of the separate names cited here. Though everyone may accept that they may legitimately be put back together again, in terms of a grasp of their characteristics, by placing them both in Class R, so named and identified. Thus *Homo* M is rightly classed with lice and trees as an organism, in the manner sketched out in the introduction, because all the named things can be said to 'live'. Science and engines are sensibly classed as inert artefacts, given these assumptions: the single word 'science' has most commonly been used in English, for more than a century, in its first dictionary sense as a body of knowledge (with 'body' used metaphorically as an assembly) fashioned so that the named species, *Homo*, can accumulate its knowledge of the single world and the single universe about it, with that singularity having been proposed and been widely accepted from the time of Newton. The single word 'engine' is used here in the special and extended sense that Engine E may be assumed to have been something made either by those called 'engineers' or by their specialist forebears; thus engines are things which are usually made and maintained in working condition by those called enginewrights in the introduction above (machines, roads, buildings, productive systems, included).

As used here, the single English word 'science' may be considered to refer to an assembly of articulated, tested and published knowledge about *Homo*'s surroundings, so that most people would accept that this comprises what *Homo* collectively claims to 'know' about such matters. The precise manner, and any observed characteristics, are problematical as to the means of articulation and test involved; and the issue of a reckless discussion by the *logographoi* of 'scientific method' and of a 'scientific rationality' of human action will be aimplified later, especially while treating further the sacralization of the idea(s) of 'science', attendant on a human vagueness about its (their) identity or identities. But artefacts tend to be defined and understood from consideration of themselves, and not from a consideration of tales told of their making. The issues of 'observation' and 'objectivity', as already raised here, ought to be readily understood by all readers of the present piece of writing who have spent a little time in school 'science labs'. And this statement may be articulated usefully in the course of any attempt, by profane means, to unlock the mysteries of science and the pieties of the scholars: Statement A, proposed and generally accepted as being a proper part of science at Time T, being a statement constructed about named aspects of the human milieu and some of their discernible characteristics, will have been assembled and tested in a particular manner. Those human beings involved will have endeavoured, as much as possible, to minimize their personal influence in the construct itself, and so in the product of their efforts. Thus, although *Homo* may well have been the forceful, wilful and meddlesome organism discussed in the introduction, this species has used its will collectively, as proposer, tester and potential user of Statement A, in the special way hinted at. Some horses have been bred to run as fast as possible. Most engines have been made to operate as cheaply as possible. The scientists and their forebears have examined and considered subjects in 'nature'; and, when they have produced representations of these subjects in the form of *logoi* of a type, they have sought to produce the maximum possible 'objectivity' in those objects (the logs themselves, usually, in recent generations, expressed on paper). It is thus that the Historians of Science have principally addressed themselves to trying to understand what *Homo* has known in the past about the human milieu, so addressing science-as-knowledge (an object made by *logographoi*).

But the often-seen designation '*Natural Science*' is as misleading as the phrase 'Scientific Method',[16] the latter used to relate to aspects of a process of inquiry, about which the Historians and Philosophers of Science tend to have had too little detailed evidence to assume the real-world existence of such a thing. For, in the reality of their inquiries and their loggings, and with the assumption of a unique world already noted here, the scientists have consistently dealt with artificial things, such as current electricity and manufactured chemicals. And, given that it has been Statement A which has been put to the test, and not any accounts and logs offered about how any human deed was done, it is reasonable to assume that the makers and testers of science and its forebears will have used all the human qualities of action that they may have had at their disposal for this making and testing. In particular, active *Homo*, observed as having been a wilful, passionate, sentient and imaginative animal, must be expected to have used all these named qualities as a part of the construction of science, whatever the characteristics of science may have been observed to be.

Here, then, are some prime candidates for consideration as being among those devices used to set out the constructed and jargon-ridden meta-world, as mentioned above, as something entirely separate from the single world and the single universe around *Homo*. Natural Science comprises an exceedingly misleading name and notion for treating what the *Cambridge Modern History* has called 'reality' in the quotation used above, since science is an abstraction and an artefact — so too it is a piece of cultural construction; furthermore, the knowledge therein logged down has not only been about natural things and events. The same *History* has discussed the possibility of a seventeenth-century generation of a 'method of discovery that would provide both explanations of nature and a rational basis for technology'.[17] But a Scientific Method is no more believable for the everyday circumstances of human activity than a Needleworking Method, given *Homo*'s observed qualities as a learner capable of active 'improvements'. Although the actions of the scientists and their forebears, including those of the natural philosophers of the seventeenth century, have often been rationalized as those of *Scientific Discovery*, this tag has been misleading about the span of time involved and the character of science. Few people now believe that (say) 'Newton's discovery of the Laws of Motion' was an act of a single instant, as the well-used phrasing implies; and, anyway, the implication that an object and an artefact (a log) was somehow found in the first instance has involved flirting with precisely the same confusion between subject and object, also between aspects of nature and culture, aspects of the past and history, which led Newton's biographer, Westfall, into a general celebration of a seventeenth-century 'mathematization of nature'.[18] That is, into assuming something must have altered, with its logging, which has been commonly been called 'nature' precisely because it has been assumed not to vary in such ways: 'The essential nature of X is . . .'

The present section has been devised to be about aspects of science, horses, engines and any other things in the human milieu whose mention may help to make a particular case on paper. Observedly, engines and horses are bulky and valuable things which can be traded; and they have the particular characteristic that, if any animals are close to them in space/time, it is advisable for those animals to be aware of the precise location of (say) Engine E and Nordic Prince. Misgrasp of this could easily lead to injury or the expiration of life for the misgrasper involved. Statement A of science, in contrast, weighs nothing in its form of an abstraction; and it weighs only a negligible amount, compared with a typical engine or a typical horse, when it has been logged on to paper for transmission to other people. In such a concrete form it would not be likely to crush a human being or trample any organism to death. But what of the issue raised about relative value?

Horses and engines can be said sensibly to be members of a wider class of thing named by economists as 'scarce goods', being goods which will alter their stock and/or condition on transfer from *Homo* M to *Homo* N. Science, however, is a 'public good' without such qualities;

and so too are the main products of the working efforts of the -ologists known as biologists, anthropologists and sociologists. Unlike the case of Engine E and Horse H, when each of these is traded between *Homo* M and *Homo* N, whether for love, money or in a barter deal, when M has informed N what M knows in the form of Statement A of science, M still knows what M knew before, whether the transfer was by what is called 'publication' in the trade, or by word of mouth; furthermore, there has been no more an alteration in the condition of Statement A than if that statement had been put to use by an engine-wright. The virtually weightless Statement A, therefore, can have no value at all in the way that scarce goods have value; and an implication of this difference, to be treated more fully later in the output of this present piece of logging (itself with no value, if only because of its author's will to see it 'published'), can be sought in the differing ways in which the importance of science and engines has been propagated by interested human groups. For Western *logographoi* of the last few generations seem to have felt a need to invent a meta-world in order to stress the importance of the thing(s) called 'science', often rendered sacred by a lack of profanation.

It has been part of the culture of the scientists and their forerunners that the inventor of Statement A should want to publish it widely — this being the nearest equivalent to the sale, by its inventor, of Engine E — as such a process is necessary for A to be agreed later to be a part of science at Time T. So the scientists and the *logographoi* have tended to ask questions such as these about the product of logging named: 'Is A as near as possible to a notional truth about the things and events named and identified in it?'; 'Is A nearer to that notional truth than rival Statements B and C about the same or related things?'; 'Does A accord well enough with the evidence from observation on which its instigators claim that it has been based?' Quite different types of question will typically have been invented, phrased and tackled regarding any testing of scarce goods such as Engine E and Nordic Prince, with this, in itself, helping to deny a core assumption of the discussants of 'technology' — that the engine-wrights, being loggers of a type like the physicists, must use working methods similar to those developed in laboratories.[19] Here are some of the types of question posed, about scarce goods in general and engines in particular: 'What use will it be to me, or to anyone else to whom I can contrive to sell it on for a profit?'; 'What does the thing do, in terms of a dynamic function?'; 'How much does it cost to operate?'; 'How much does it cost to maintain, when it is operating or not?'; 'Who owns it?'; 'How much does it weigh, for purposes of installation and/or use?'; 'How easy is it to transport, whether under its own power or under the power of other engines?'; 'Is it dangerous for a human being to be close by it when it is operating?'; 'If I try to share a particular point in space/time with it, will it be likely to kill me?'

Science is conceived here to be a provisional body of knowledge of aspects of the single world and universe about *Homo*, with the consequence that *Exact Science* can only be a misleading notion about what the quotation from the *Cambridge Modern History* has called reality. If the dissecting and sacrilegious approach of the scientists (that of temple robbing) is used to treat uses of the single tag, 'science', it is apparent that a use of this tag to refer to a science-as-an-activity as well as a science-as-knowledge, has been reckless. Simply because the latter of the two can be assumed reasonably, in the manner of matching the results of observation with abstractions put on to paper, to have been particularly orderly and rational in that way; and simply because the scientists have claimed to have discerned laws and principles and other aspects of regularity in 'nature' (that which they have chosen to study); it makes no good sense to have tried to make science-as-an-activity appear rational and orderly to match. Accordingly, the idea of a *Rational Science* is not believable as related to a real-world *Scientific Rationality* of *Homo*.[20] Furthermore, the issues of ownership, bulk, and so on raised here can be used to help deny the utility of the term *Applied Science* for understanding observable, real-world things and events.

Because of the types of question posed about Statement A, whose inventors and publishers may wish to have it classed with science, there is no conceivable way in which it can be filed within (say) 'physics', so implying that it cannot be a part of 'chemistry' concurrently. The physicists cannot lay claim to own it, any more than *Homo* M, who may own Nordic Prince, cannot own Statement A either. Newton's Laws of Motion may well have been the most applied-to-use statements of science — though no one can know that in a tested way. So one person's piece of *Basic Science* or *Pure Science* could turn out to be identical with what another person has cared to dub as part of 'applied science' — this, in particular conditions, for the public good called 'science', that any use of it by engine-wrights or anyone else makes no difference to its stock or condition before use. This is, as well, in the particular conditions for the actions of the scientists and the engine-wrights, which are to be considered more fully in the following section and have been sketched out already, that there is no discernible way for *Homo* to act 'scientifically' — this latter to be recognized in the activity, as opposed to its product. Scientific method has never been a name referring to anything observed in the single world about *Homo*, apart from in the accounts made by the *logographoi*; accordingly, Popper's title, *The Logic of Scientific Discovery*,[21] has been doubly misleading, since no one has been able to identify any *logoi* consistently at the heart of the scientists' inventive actions. Besides that, no one can sensibly claim to know by how much science-as-knowledge has been used in typical human activity. So the hunt must go on to try to uncover any discernible mysteries of active 'Homocraft', as this has gone on outside the playhouse theatrics of the scholars' meta-world of paper-borne abstraction. Could we conceivably stand in need of the services of M. Poirot to puzzle this one out?

ACTIVE *HOMO*: MACHINE-USER

In the previous sections an idea of science has been developed in the normal manner in which our species uses words and builds up definitions. Science is the result of a logging, being an object compared with the subject(s) of the scientists' special concerns in their inquiries; and an operation of the principle of objectivity has been arranged so as to minimize any influence of the details of inquiry on science itself, which will be called Hamlet's science here, for purposes of identification. So what about the character of any real-life skills of the scientists, the mark of which have been shaded out of Hamlet's science consistently? What of such skills, in the particular circumstances, wherein the Western *logographoi* have shown great awe for the thing(s) they have chosen to call 'science', if only through their feeble attempts to tie it (them) down in profane definition? What can be said about these skills, given the *Cambridge Modern History*'s eccentric use of a phrase, to have claimed 'secularisation [to have been] primarily . . . the process of change . . . to its [reality's] interpretation in terms which are essentially natural and focussed on this world'? In fact, such 'terms' used have consistently been constructed things (force, structure, cause, law, principle, oxygen), which no one has been expected to find in natural conditions, and Hamlet's science has been riddled through with the use of metaphor in order that individuals may best make their case in their loggings and their logs.

The case introduced above about an alleged and special 'rationality of science', as related to science-as-an-activity, can be amplified here with a development of the earlier argument that the character of artefacts should be sought in themselves, rather than in accounts put about concerning how they may have been made. And this argument can be fed through later to an amplification of the case sketched out in the introduction above, about how a nineteenth-century technical 'afterdom' cannot be sustained as an idea useful for understanding aspects of the past, as opposed to understanding aspects of the construction of histories.[22]

The attribution of rational conduct mentioned here refers to aspects of human dynamics, with the particular word 'dynamics' used in the same sense as when historians have discussed 'Newton's unified dynamics', often said to have been a seminal set of statements of what has been called the Scientific Revolution latterly. That was a set of statements explicitly about what are now called 'inert bodies in motion', articulated as part of what was often called 'the mechanical philosophy' at the time of Newton's first articulation of such things. The present section will concentrate on some observed aspects of human dynamics, and so too of wilful and forceful behaviour, as this particular group of animals has, often in collectives, engaged in a range of actions in its constructive and meddling way – making and operating engines, professing its various faiths, making and testing Hamlet's science and its forerunners, waging wars, trading, crusading, running its polities, etc. *Homo* will be considered first as an active individual, with a notional boundary thrown around this animal organism, and as a long-standing machine user. The notions of 'acting' and 'performing' involved are not those of the theatre and the playhouse, where some say that humans try to simulate human actions elsewhere. This has to do with what the *Cambridge Modern History* has conceived as a 'reality' outside the written page and outside the playhouse.

The human animal, it is true, may be considered as a fairly static thing, rather as engines in general and machines (including tools) in particular can be so considered, broadly in the terms of costly, tradeable goods as outlined above. Thus, the particular organism named here may be depicted, if the depicter wishes to do so, in various modes — sleeping, at prayer, sitting in a study, trying to appear wise, ill and 'indisposed' in bed, pondering relationships within the human milieu and so forth. But none of these modes helps enlighten anyone about the formulation of the sketch of *Homo* offered above, as a forceful, 'disposed'-to-action, apparent magician who has influenced the planet Earth strongly. This is *Homo*, which has started, and mainly conducted, two conflicts in its most 'civilized' areas in the twentieth century, with these disputes commonly known as 'world wars'. This is *Homo*, whose characteristic human behaviour of maiming, torturing and killing those of its own species, has been commonly known as 'bestial' by type. (Thoughtful lice might wish to call such conduct characteristically human.)

As such an agent of alteration of its milieu, as an impinger on many things around it, and as an animal species which might be said to have shown a marked flair for such things, *Homo* is best conceived (taken together as) as the perennial constructor (arranger of things together) in a wide range of varied enterprises. This has been so in the general manner of assembly (putting things together), along with the transportation (moving across) of things — raw materials, components (things placed together), people, finished products. This overall exercise has involved much contrivance (artifice), as anyone who has taken part will confirm; and constantly it has taken place in broadly experimental modes, in the sense that detailed outcomes have not been completely (filled together) predictable before the events of action.[23]

While there have been many human commentators on such matters who have endeavoured to discern the parts played by named components of *Homo* — perhaps heart versus head, body and soul, mind and matter — it is reasonable, for present purposes, to place a notional boundary around the whole individual human being so as to try to make some sense of skilful behaviour on paper. Individual *Homo* can be considered as having acted dynamically as a whole organism in forceful ways. And the tendency of latter-day Western *logographoi* to try to analyse the actions of real-world, ostensibly logging 'technologists', can be seen at once to have been very eccentric, and also to have laid a trap for those who would then go on to debate 'the relationship between science and technology' and how this may have changed over time. 'Not a very rational designation. See me! Teacher.'

That said, any observant and imaginative louse or dog might have been expected, if it were able to articulate speech, to have joined those members of the species *Homo* who had let their curiosity rule their conceit, to agree a formulation of words broadly along these lines: *Homo* has been a maker and a doer for as far back in time as it is possible to grasp evidence of this species's stay on the single planet, Earth. As such, it has logged things, graphed things and made engines. It has produced and operated legal systems and polities. It has made Hamlet's science and Works of Art, as well as a whole range of the things hinted at above. Moving things about the planet (bank transfers, transmission of information, transported goods and people, metaphors in speech and writing), and arranging things together skilfully (moulding, shaping, cutting, conferring, making syntheses and fields and languages, devising human collectives and religious confessions), *Homo* has operated in a variety of specific ways. And *Homo* has made use of a range of machines (pens, hoes, lathes, locomotives, cars, prime movers, axes, swords, ploughs) in these separate resolves. (This involves defining tools as machines, as engineers tend to do.)

As members of this forceful species, medical doctors have used similar devices to those which other people have used in warfare (say) to rearrange the configurations of human organisms to preferred conditions. Civilian groups concerned with 'management' have borrowed techniques invented for running armies equipped with 'infernal machines' of death and destruction. Such borrowings and purloinings have been virtually endless. They have been in fact typical of the enterprises in which individual members of our species will tend to claim a wish to 'get their act together', in which individuals and collectives of *Homo* have been ever fearful that 'it may all fall apart' or 'it will come apart at the seams'. 'The time is out of joint', bemoaned a fictional Prince Hamlet, facing catastrophe.

So as to deal effectively with these types of affair in articulate speech and writing, *Homo* has selected for use the word 'technique' from an Old Hellenic root, *techne*, normally given in today's English as 'skill' or 'art'. The English word 'artistry', as used today, can – along with the English words and phrases 'artisan', 'the Arts', 'verbal articulation', 'artful conduct', 'a journal article' – be linked through to Latin and earlier roots, with words indicating the jointing of things together: *artus* is a Latin word usually translated as 'a joint' in current English; and the English idea of 'arms' is that of weapons metaphorically jointed to the owner or carrier, for that individual's protection, as if they were parts of the individual member of the species *Homo*. Tools have typically been made to act in the manner of machines, as will be shown below. So, for anyone who may wish to bear in mind what the ill-named 'technologist' may have been up to down the years, making and using bulky and scarce goods, which have such value as may be indicated in a market, it is worth noting that one designation of a carpenter is a 'joiner'. The accent should be, for comprehension of the details of 'Homocraft', on the real-life, off-stage actions of Prince Hamlet, conceived as what he was made to call an 'enginer' in the play written about him, and conceived without such building blocks of a meta-reality as Scientific Method and Science-Based Technology and Industrial Revolution.

Machines are among the wider class of things discussed above as engines, the main specialist concerns of the engine-wrights. A machine has often been defined as a device made to work with a particular dynamic function: it transmits a force or energy which has been applied to it by an external agency at Point P, to be delivered at another, Point Q, elsewhere, in the form of a force or energy to an agent outside the machine itself. Thus conceived, machines are force transmitters, in their typical modes of action, as used by a forceful species, *Homo*. A tool (a lever, or a sword, or a saw) transmits internally, in the same general way as a powered machine (one powered by another agency besides *Homo*); and an electronic computer, another machine, appears to have transmitted information internally, through having converted this into energy. As with the case of the model of dynamic *Homo* proposed here, machines are hereby conceived as actors with a notional boundary thrown around each of them, so that any propagated reasons for *Homo* M's having exercised (say) Dynamic Function F need not be considered here. Furthermore, it is not necessary to try to speculate here what may have been going on within the boundaries proposed. What needs to be consideed, in contrast, is that Points P and Q are located on the boundary of Machine X, with the last-named considered as a dynamic system in operation; also that *Homo* M, its user, will have tended to divide out parts of shared tasks and functions between M and Machine X.

'The division of labour' is a notion that has become a commonplace in recent generations of the species, *Homo*, though many of that species may wish to assume that this 'principle' was put into active practice long before it was articulated in those terms. In such a way, *Homo* M and *Homo* N have shared out dynamic tasks and functions, so that each will tend to specialize in what the individual seems to be better at doing. Using the same analytical (literally, loosening up) and dissecting notion, in the manner of sacrilege and critical disbelief discussed above, it is apparent that *Homo* M and Machine X will have divided things out between them in the same general way of specialization. So it is worth stressing here that one of the few observed characteristics of machines, besides those raised above (that they are scarce, valuable and bulky force transmitters and energy transmitters) concerns their dynamic predictability. For Machine X, when it is working normally, pre-diction and post-diction of its actions tend to be identical and articulable in some form by *Homo* on to paper — in algebra, in words, as a set of characteristic curves, etc.

Accordingly, *Homo* M, by far the more adaptable of the two labour dividers of the pair of them, has been forced to deal with those parts of dynamic functions which are the least well predicted and the least easily articulated on to paper or to its local equivalent of time and place. This forceful animal, *Homo*, has thus been forced itself into particular modes of action, in the sense of being obliged so to act if it wishes to succeed as a real-world machine user.

The successful operation and completion of virtually any dynamic function with which *Homo* has been involved has required making judgements about the future and about conditions to be expected in their details. This has been the case for relatively simple affairs such as crossing the road, to more complex ones such as some of those mentioned or hinted at here: engine building, science construction, 'running' countries and workplaces. Since only a reckless fool will want to be found claiming that the future is 'known' – and since everyone knows well, too, that many aspects of the past are always problematical, in the sense of knowing and understanding its particulars – it follows that Machine X and *Homo* M will typically have faced together components of risk and the relatively unknown. This has been the case, as well, for collectives of *Homo* and *Machina*, wherein activity and effort will be expected to have been divided up in the same general manner.

Accordingly, real-life and specifically human dynamics, and any personal skills deployed by *Homo* in its conduct, must be considered as having centred strongly on risk-taking and treating the unknown. So these have not been something readily dealt with by the *logographoi* and in print, within a Western intellectual tradition given by statements such as this: 'The Ionian philosophers [of the sixth and fifth centuries BC] were doing something unprecedented: they were assuming that the universe is an intelligible whole and that through rational inquiries men might discover the general principles that govern it.'[24] On the contrary, the perception of an imaginary and imaginative louse or a pig is, in the general manner of the notion sketched out in the introduction, more reliable about active 'Homocraft' in this universe: as apparent magicians, human beings have not typically acted intelligibly, in the sense of discernible 'principles' playing out a part of 'governing' action taken up. Furthermore, the analytical devices as invented by *Homo* at the time of Boyle and Descartes and Newton, and as attributed then to 'the mechanical philosophy', have similarly provided unreliable guides to treating active *Homo*. This has been the case if only because the predictability of machine dynamics has been quite different from the lack of predictability of human dynamics, however vocal and pious have been the scholars' calls made about *Industrial Revolution* and its subsequent *Science-Based Technology*.

Considered thus, an active 'Homocraft' is, and always has been, something difficult to rationalize on to paper in the form of meaningful generalizations. *Homo*, in sharing actions with machines, has simply taken up the less regular, less norm-affected parts of these actions in the course of trying to grapple with and reorder the particular.

The particular question cited in the Introduction from Musson and Robinson, about technical conditions in an ill-tagged Industrial Revolution, is misleading and unanswerable because, in the absence of any discernible 'scientific basis' for it, the exercise of technical skill has always had some type of 'practical tinkering' at its core. So the seventeenth-century vision, as cited above from the *Cambridge Modern History* about 'a method of discovery that would provide both explanations of nature and a rational basis for technology', has been very misleading in two separable ways. Typically, opportunistic *Homo*, as observed to show personal flair, technique and some artistry in its forceful and constructive operations, is not well or usefully considered as having been a particularly 'methodical' organism, with its actions 'based' on or 'governed' by any externally generated precepts. Furthermore, a real-life Hamlet, striving to deal successfully with bulky, valuable engines which have to work in regimes where cost is crucial, must be expected to have developed for use a different range of skills from (say) Horatio, a scientist dealing with abstractions with no market value, and often operating nowadays in the controlled regime of a laboratory. So now to some of the features of how Hamlet and Horatio were drifting apart from each other physically even in the nineteenth century, as signalled in the introduction here concerning 'the faith of 1874' and a need felt by some people to try to 'sell' an indistinct 'science'.

1874, DIVERGENCE AND ALL THAT

Earlier comment here has indicated how, in the conditions of the abstract meta-world invented by the *logographoi* of the West, (an) awesome and ill-defined thing(s) called 'science' has (have) been made to appear magically powerful in the real world. Science-as-knowledge (Hamlet's science, here) is an inert body of knowledge, normally lying inactively in print, incapable of a dynamic function unless harnessed by some other agency. Yet, by designating something else with the same name, science-as-an-activity, the scholars seem to have breathed life into inert science-as-knowledge, appearing to have awarded it active functions like its namesake. 'Science builds a new world.' 'The activity of science has dramatically transformed society; since 1850 applied science has become the basis of the means of economic production in Europe' (*Cambridge History*).[25] In such a way, the historians' pathetic fallacy can be seen in full operation, with Hamlet's science appearing to be alive, while Hamlet has been stripped of lively skills to match. *Homo*, whom we all meet daily, spirited and opinionated, is scarcely to be seen.

The two manifestations of 'science' named here, however, do not represent the end of the particular tale of fudge and lack of profanation. There has been as well the 'two cultures' notion of science-as-a-culture, which seems to have been confirmed from the name given to the main museum of engineering products in the country where the present paper has been written, from the name given to the Science Museum in South Kensington in England. And there has been, too, Cardwell's notion of science-as-an-institution in his version of the thesis of what the *Britannica* has called 'the convergence between science and technology in the 19th and 20th centuries' – this, as dealt with as well in the volume of the *Cambridge Economic History* cited from above, whose editors seem to have been under the misapprehension that the USA and Japan are part of Europe.[26] But, before returning to that matter, it may be worth listing a few more components of the scholars' meta-world as well as of real-world joinery.

The Scientific Revolution is a phrase which has only become commonly known in the last few genertions, and concerns various events celebrated in and about what has been called an Early Modern Europe. This rather slowly wrought shift in the accepted world-view is sometimes said to have centred on Newton's published synthesis of 1687 about 'the revolutions of the heavenly bodies' together with the motions of inert bodies terrestrially, with the use of a machine metaphor to explain both sets of motion. And the Revolution is often said to have provided suitably modern means for both expressing the subject of 'nature' on to paper as an object and for active *Homo* to understand how to act 'scientifically'. Yet even the 'revolutions' of the one phrase have been of quite a different type from those of the second. The first phrase involves conspicuous overdramatization, as with a use of 'scientific discovery', of time-scales elasped. And no norms were ever established, in an aftermath of change, for either of the main varieties of 'science' discussed here. In particular, when faced with the special problems of making logs of their chosen subjects of concern, the scientists and their forerunners have invented, and used, any devices which have seemed to suit the particular process of what has been called objectification above.[27]

The Industrial Revolution has usually been adopted as an abstraction to refer to a set of events, scarcely so tagged at the time, which started in the mid-eighteenth century in Britain, and are thought to have changed the rest of the world in some ways with their subsequent spreading elsewhere.[28] So, manufacturing has been placed into focus from the name adopted; and Watt steam-engines, with their consequent machine-paced factory regimes, have been much in evidence in print, on this and associated topics. Yet, at the time when Britain has been thought to have first become an 'industrial nation', there were probably more domestic servants at work then factory workers. By the end of the nineteenth century, and with what historians now sometimes call the Second Industrial Revolution, a use of electric motors was starting to overturn any dictates of the centralization of 'production' via Wattery. And, anyway, there has been no experience of a major nation with its 'product' stemming mainly from manufacturing. Although the economic historians have acted as pious gatekeepers to protect their favourite idea of Revolution, even the comments of a single paragraph of prose such as the present one can be used to question the naming, the imagery and the cognition involved about allegedly real-world events.[29]

Science-Based Industry, one of the volumes of the *Cambridge History* dealing with the nineteenth century, has distinguished 'scientific' from 'technological' types of invention.[30] There have been many other propositions in print about the advent of a 'science-based technology', as noted and implied above. And there has been a whole debate about how the 'practical men' of 'industry' had their way of working overturned at some time in the past. But 'humanistic' *logographoi*, who had struggled out of their studies so as to take a close look at near-colleagues, the 'scientists', at work, would probably have found that even these folk have had to be very practical people in their own ways. That is, in their own modes of dealing with the particular, in circumstances where no one would know what to do next, if exhorted from outside to 'be scientific'; and there has not been anything akin to (say) discernible 'grounds for divorce', to generate expressly founded-on-knowledge, science-based modes of working.

As for the practice of joinery, which has disappeared less than most of the ill-named 'technology' behind the opaque walls of the various things called 'industry' (most workplaces), anyone who cares to do so may contrive to observe a carpenter making a kitchen cupboard in the observer's own household. That is, making a scarce good in the sense outlined above, something bulky and valuable to its owner and devised with particular utility in mind. That is, as well, making something distinctly unlike the logs and the abstractions of the scientists, for which claims made about potential utility have had no bearing on tests made about their adequacy. Sometimes using an explicit result of graphing and logging called a 'plan', and sometimes not so, with any such-named thing being analogous to 'design drawings and specifications' in engineering, the joiner may be observed doing a wide range of things: moving components (things placed together) about; shaping them and the places where they should fit; going out to get materials for use; assembling things at first; altering the details of such assembly as a result of such testing; cutting and measuring things; taking a break for sustenance and relaxation; trying to imagine what a future configuration would look like, etc. In such a general manner, a specialist carpenter will be seen using tools and machines in the manner of predictable machine activity, with the aim to reorder a part of the future cupboard-owner's house, and by 'engineering' solutions personally to what in particular had been judged to have been 'problematical'. Thus, a 'practised' *Homo* will generally have been expected to be more competent actively than an amateur; the former will be expected to operate in a less overtly planned mode than the latter; and, with increased personal competence, any practitioner will have been expected to be found operating in a progressively less overtly 'science-based' manner. *Homo* will be more likely to claim, with justification, 'I know what to do', when advice and information is offered by other people, the more specialized and experienced in specialism he or she has become. 'Progress achieved!'

Now, back to what has been called 'the faith of 1874' above, so to invoke the phrasing chosen by a president of the Royal Society of London about 'the faith of 1660', that being the year of the body's foundation.[31] Cardwell's theme, in a book entitled *Technology, Science and History*, has addressed what he has discerned as four phases in 'the development of technics and its evolution into *technology*, which we [*sic*] shall briefly define . . . as technics based on science . . .'[32] — this being a definition which implies the metaphor, in engineering, of a column and its necessary and special base, which definition readers of the present paper may wish to question. Although the theme of Cardwell's book has broadly been that of Forbes's *The Conquest of Nature*, to log down details of a convergence or a coming together, with 'science' cropping up (at times, at least) as 'a progressive, cumulative activity' by *Homo*; with a logging of a 'third phase in the development of technology' (apparently, an activity involved with engine-making), the focus seems later to have changed in this book, which is ostensibly about single things called 'science' and 'technology'. So, with a notional 'fourth phase' in the past, as some fully fledged 'technology' came into being, with both 'the establishment of industrial research laboratories' adopted as a symbol of convergence and the new technical importance of Germany, the USA and Japan (all of which countries the editors of the *Cambridge Economic History* have located in Europe) in the mid-nineteenth century — abracadabra!, we have even more change in the frame: 'The conspicuous feature of the period was the rapid convergence of science and technology considered as social institutions; the virtual completion of the Baconian programme.'[33] Science and technology seem to have changed their identities, with the phasing out of the earlier 'technics'! But did anything significant occur in the real world, as opposed to a bookish meta-reality, in this past era of increased specialization, at a time of what some have called 'the professionalization of science',[34] and with a consequent programme of propagation of the great potential utility of the thing(s) called 'science'?

In the particular conditions of the real world about *Homo* there was little overt and general discussion about anything at all called 'science' until about the nineteenth century. No one was called a 'scientist' until then. There were almost no people living who derived a livelihood from trying to log 'nature'. Newton and Galileo had been professors of mathematics in their years of greatest achievement. Boyle and Descartes were gentlemen-inquirers. William Harvey was a physician, one of whose patients was Bacon. The Royal Society of London did not close its membership, to include only those whom we would call 'practising scientists' today, until the mid-nineteenth century. It was barely possible to obtain a university degree in science anywhere before then. There was no significant and collected group of people anywhere before then whose members might have had the self-interest to boost the thing(s) called 'science', and to invent confusing abstractions such as science-as-a-culture, science-as-an-activity or science-as-an-institution. No one had felt the need to debate any 'relationship' between 'science' and anything else, primarily because there was nothing in existence whose 'convergence' to anything else seemed to need to be tracked down. 'In 1869 most men of science in England were amateurs . . . It was only at the turn of the century that a class of professionasl scientists, comparable to the one that had existed in Germany for forty years or so, began to appear in Britain' (Cardwell, providing good evidence to deny his own thesis of 'stages' in 'technology').[35] Precisely so; and it has only been from the nineteenth century and on, with its observable divergence of people called 'scientists' by occupational specialism, as these folk have split off to work in places for which the Cavendish may be used as a symbol here — so splitting off from everyone else, in the normal manner of the division of human labour and effort — that the meta-world discussed here has been invented and propagated by the scholars, *logographoi*.

The present piece called 'Hamlet without the Prince' has hinted obliquely at Bacon's notion of the Idols of the Theatre, while stressing a dramatic other-worldliness in the boosting of a valueless science; and so it has referred to Bacon's argument about false presuppositions in inquiry. 'Hamlet' has referred as well to a notional Prince of Denmark, who might be imagined by readers going about a real-life, off-stage activity of 'Homocraft' — addressing, in the skilful and opportunistic way of normal human action, the particulars of things judged to have been problematical, and so in need of some reordering. That is to say, there are certain things said, at the particular time and place concerned, to be worthy of an attempted solution, with that solution later perhaps to have been rationalized (in the verbal chic of the time and place) as having been 'progressive' or 'humane' or 'scientific' of the members of the forceful species *Homo* who were the instigators of change.[36]

The convergence proposition of the *logographoi*, as criticized here, can be said to have been a piece of historians' fancy, because, with its insistence on making out the engine-wright to have been a logger and grapher, it has not focused on the details of 'Homocraft'. All the while, an apparently pathetic, living 'science' has been held in great awe; and it has been celebrated recklessly as part of both the Western intellectual tradition of claiming future intelligibility in the world and the faith of 1874 mentioned above. A tale told of drama in 'technology' has been theatrical, but not believable. 'There remains Galileo's role in the unification of science, and I have already given away the secret', A.R. Hall has written in a book entitled *Galileo: Man of Science*. 'The mathematization of nature itself entails a certain unity in the world-view; it is imposed by the opinion that there is no knowledge more fundamental than that of the mathematical physicist. I have already attributed such an opinion to Galileo.'[37] *Pace* Galileo and Hall, nature was not mathematized with the construction of part of its log in such terms. Visions of a unified world-view, as put together from the Scientific Revolution to treat inert things, have been mischievously put to use for logging wilful *Homo*; but they have provided no better guide to cognition than the predictability of machine action has provided a meaningful metaphor for real-life human activity. The meta-world wrought on to paper, with its confusions between aspects of the past and of histories, has only been part of a playhouse show of such favourite things as a 'mathematical physics' puffed up as being 'basic' or 'fundamental' to other things, while no one has tackled adequately the issues of skill and working technique.[38] 'Scientific basis, my foot!'

Notes and References

1. 'The History of Technology', in *Encyclopaedia Britannica* (Chicago, 1992), Vol. 28, 452.

2. A. Christie, *The Pale Horse* (London, 1961), Fontana edn, 95.

3. R.J. Forbes, *The Conquest of Nature* (London, 1968), Pelican paperback edn, 113. A bizarre effect of the odd tradition in naming can be seen in the content of the *Britannica* piece, *op. cit.* (1). From the definition given in its first sentence, the article should have comprised a history of 'the systematic study of techniques for making and doing things', a study of human beings studying things; but, in fact, it has addressed mainly the making and its effects.

4. *Oxford Classical Dictionary* (Oxford, 1970), 521.

5. David S. Landes, 'Technological Change and Development in Western Europe, 1750–1914', in *The Cambridge Economic History of Europe* (Cambridge, 1965), Vol. 6, 274–601.

6. *Ibid.*, 550. The book named (Cambridge, 1969).

7. *Ibid.*, 550, 551; D.S.L. Cardwell, *Technology, Science and History* (London, 1972), 2.

8. Landes, *op. cit.* (5), 550, 551.

9. *Ibid.*, 550.

10. A.E. Musson and E. Robinson, *Science and Technology in the Industrial Revolution* (Manchester, 1969), 1, 65.

11. Christie, *op. cit.* (2), 223. In this story occult practices were set up as an elaborate cover for systematic murdering.

12. Peter Burke, 'Religion and Secularisation', in *The New Cambridge Modern History* (Cambridge, 1979), Vol. 13, 293–317, p. 294.

13. This has been cited as one of the three main concerns of the contributors to *Technology and Culture*, in J.M. Staudenmaier, *Technology's Storytellers* (Cambridge, Mass., paperback edn, 1989, Ch. 3, esp. p. 83. Concerning an indistinct 'technology' — whereas this single word seems to have been tagged mainly to an activity in Ch. 3, in the previous chapter (expressly on Emerging Technology, another main concern of the journal), hey presto!, the same word seems to have been tagged to engines and the like concerning 'The Process of Technological Change' (pp. 40–61).

14. To some extent, the present case is consistent with that of Walter G. Vincenti, as expressed in *What Engineers Know and How They Know It: Analytical*

Studies from Aeronautical History (Baltimore, 1990) and elsewhere. Yet the idea of humans 'knowing' something has always implied an ability by the 'knower' to be able to articulate 'knowledge'; and the tendency to class 'technical skill' as a part of 'technological knowledge', as discussed in Staudenmaier, *op. cit.* (13), 114–20, is rejected in the present article. Skill is treated here as something different from knowledge by reason of its essential character.

15. Cf. F. Bacon, *Novum Organum* (London, 1620), where the Idols of the Theatre were invented to refer to false presuppositions in studying nature.

16. Phrases italicized thus are here assumed to have been, like Science-Based Technology, part of the build-up of the criticized meta-world. Scientific Method has been treated as a transcendental entity in M. Fores, 'Homo et Machina', *History of Science*, 1994, 23: 63–87.

17. A.C. Crombie and M.A. Hoskin, 'The Scientific Movement and Its Influence 1610–50', in *The New Cambridge Modern History* (Cambridge, 1970), Vol. 4, 132–168, p. 133.

18. R.S. Westfall, *Never at Rest* (Cambridge, 1980), 14.

19. Cf. the precise wording about 'method' and 'rationality', of the quotation above from Vol. 4 of the *Cambridge Modern History*, where working similarity has been assumed.

20. Cf. the critique in M. Fores, 'A Cambridge View of Modernity', *Journal of Historical Sociology*, 1992, 5: 209–233, pp. 220–4.

21. K. Popper, *The Logic of Scientific Discovery* (London, 1959).

22. The idea of the Theatre of the Bourgeoisie may help on this, as in M. Fores, 'Newton on a Horse', *History of Science*, 1985, 23: 351–378, pp. 368–70.

23. This model has been developed more fully in Fores, *op. cit.* (16).

24. 'The Study of History', in *Encyclopaedia Britannica* (Chicago, 1992), Vol. 20, 559–632, p. 559.

25. P.M. Heimann, 'The Scientific Revolutions', in *The New Cambridge Modern History*, *op. cit.* (12), 248–270, p. 248. This is the first sentence of the chapter.

26. Quote from *op. cit.*. (1), 482. There is a chapter, or a part of one, on Industrialization in each of these countries in *op. cit.* (5).

27. Compare my piece on 'A Cambridge View', *op. cit.* (20), 220–4.

28. Thus, the volume after the *Cambridge Economic History*'s treatment of the Industrial Revolution, *op. cit.* (5), has addressed *The Industrial Economies* (Cambridge, 1978), those having experienced 'industrialization'.

29. Compare my piece on 'A Cambridge View', *op. cit.* (20), 224–9.

30. A.R. Hall, 'The Scientific Movement and its Influence on Thought and Material Development', in *The New Cambridge Modern History* (Cambridge, 1960), Vol. 10, 49–75, p. 49. My piece, 'Newton on a Horse', *op. cit.* (22) was written in criticism of the synthesis on 'science' and 'technology' of this *History*.

31. 'According to any sane assessment the faith of 1660 has been fulfilled, and in the Tercentenary celebrations we pay tribute' to earlier members. Sir Cyril Hinshelwood, 'Improvement of Natural Knowledge', in *The Royal Society Tercentenary* (London, *The Times*, 1961), 14.

32. Quotation from Cardwell, *op. cit.* (7), 2. Cardwell's emphasis.

33. *Ibid.*, 213–15. Compare Forbes, *op. cit.* (3), Part 1, esp. p. 42.

34. Cf. Heimann, *op. cit.* (25), 248, on 'the emergence of science as a socially-organised activity' at that time.

35. Cardwell, *op. cit.* (7), 216, 217.

36. Thus the present paper rejects Landes's idea of a nineteenth-century shift in practice, which involved a tendency 'to diminish reliance on the

traditional combination of empiricism and common sense and impose a more scientific approach', *op. cit.* (5), 551.

37. A.R. Hall, 'The significance of Galileo's Thought for the History of Science', in E. McMullin, *Galileo: Man of Science* (New York, 1967), 77.

38. Readers might wish to note that the idea of skill raised here, and in the title of the present paper, has been that of common usage and the normal first definitions of English dictionaries, as related directly to human abilities in action. This is a different idea from that of Staudenmaier, *op. cit.* (13), 116: 'those [*Technology and Culture*] authors who accept the design-ambient character of technological knowledge necessarily imply that engineering theory and technical skill are two irreducibly distinct components of all technological knowledge'. See also note 14 above. Compare, as well, the fourth section, on Human Activity, of M. Fores, 'Constructed Science and the Seventeenth Century "Revolution"', *History of Science*, 1984, 22: 217–244, pp. 221, 235–241.

History of Technology at the German Mining Museum, Bochum

WERNER KROKER

ABSTRACT

In 1930, after a considerable period of consideration and discussion, a 'Historical Mining Museum' was founded in Bochum, then the centre of the Ruhr coal-mining district. Today, this 'German Mining Museum' (Deutsches Bergbau-Museum) is clearly a world leader, having provided inspiration and ideas for similar institutions in various parts of the world. It ranks among the most frequently visited museums in the country. Many of its some 400,000 visitors a year come from abroad, which is the more remarkable since Bochum is not regarded as a 'must' in international tourism. But it is not only the modern replica coal- and iron-ore mine, having been constructed as close to reality as possible by the museum's miners under the premises, that attracts people. The museum's reputation derives from the large variety and multiplicity of times in which the history of technological development in all branches of the mining industry throughout many countries and ages is presented.

THE MUSEUM, ITS ORIGINS AND FUNCTION

Bochum, although there is no longer any working coal-mine within its boundaries, was one of the centres of European coal production for more than a century.

It is still an industrial town, however, with a population about the same as the annual number of visitors to the museum. The steel industry remains important to the area despite its present structural economic problems, while since the 1960s automobile production has given work to thousands of former miners after the closing down of some twenty mines within the city area. As far as the cultural environment is concerned, the Ruhr University, with more than 30,000 students and one of the leading new universities in the country, has a notable influence – as it does also on the economic and social life.

Nevertheless, since 1973 the new landmark of the town is a pit

Figure 1 Aerial view showing the winding gear on top of the German Mining Museum.

head-frame winding gear, 68 m high, which, dismantled at an abandoned coal mine in neighbouring Dortmund, has been re-erected over the museum's main building. Moreover, this rather expensive project from 1973 can be regarded as a symbolic act that initiated a series of further remarkable and successful projects to preserve the industrial heritage of North Rhine-Westphalia.[1] It was carried out at the time that the museum was itself actively involved in preservation work on industrial monuments, e.g. by establishing the Mining History Trail in one of the earliest regions of Ruhr coal-mining with the Historic Miners' Prayer House as its centre,[2] began its nation-wide work of documenting work of industrial monuments[3] and organized the Second International Congress on the Conservation of Industrial Monuments which led to the foundation of TICCIH,[4] the International Committee for the Conservation of the Industrial Heritage.

The origins of the museum can be traced back to the material contributed by the Prussian mining industry to the World Exhibition in Paris in 1867. Its main aim was to popularize the achievements in the most important coal- and ore-mining districts in Germany. The Prussian government, as well as the companies, provided a sound impression of the high technological standards and the product quality which had been gained during the previous years of industrialization as a result of liberal economics. After the fair, the exhibits were moved to Berlin in order to be incorporated into the Museum of Mining and Metallurgy which was established there in the following year and which thereafter became a part of the Mining Academy in the Prussian capital.

However, during the merger of the Academy with the Technical University in 1917 this museum was dissolved.[5] But some of its exhibits were to be preserved in the Ruhr district, where plans took shape to make historical demonstration material in the field of mining permanently accessible to as many people as possible. This idea was promoted by Westfälische Berggewerkschaftskasse (WBK), a joint organization of the Ruhr coal-mining companies established in Bochum since 1864. As training and teaching young miners in its privately run schools had been one of its main mandates since its beginnings, this 'Westphalian Mining Fund' was in possession of many technical models and other didactic equipment. Over the years many of these devices had become obsolete, but at the same time they had turned out to be important historical objects themselves, representing different periods in the development of coal-mining technology. They were, therefore, too precious to be thrown away.

As a result, in 1930, WBK and the municipality of Bochum signed a contract to establish a 'Historical Mining Museum' and to finance it jointly. The municipality, being aware that this institution would contribute positively to the city's cultural life and thus polish up its image as a workers' town, agreed to pay half of the maintenance costs in the future. Thus, it is still the first and only museum in Germany maintained by a special branch of industry in order to show its historical development and its current technological state, and coal-mining is still the only branch of German industry to have established such an institution.

Today, the 'German Mining Mueum', its official name since 1976, is still run as a department of the mining industry organization, nowadays the DeutscheMontanTechnologie (DMT), although in the meantime it has been put onto a totally new budgetary basis. After it had grown too large and turned out to be too expensive for the two founding partners they reduced their contribution to one-quarter of the annual budget each, the federal government in Bonn and the government of North Rhine-Westphalia in Düsseldorf jointly contributing to the costs of research work.

Actually, mainly because of the extension of research work as a prominent mandate of the museum, this decision was of decisive importance for its future existence since neither the municipality nor the coal-mining industry could afford the large sums needed – altogether, DM 12 million in 1993. As the museum's tasks have expanded, the number of staff members has increased to more than 120 during the past two decades. The museum's buildings have grown in number too: originally accommodated on the premises of the former slaughterhouse of the town, since 1973 the museum has been entirely rehoused in a complex of purpose-built accommodation; the buildings cover almost 25,000 m^2, about half of the space being used for the museum's exhibition area.

As a matter of fact, what may be regarded as even more important are the almost three kilometres of galleries in the demonstration mine, 15–20 m below the surface buildings. Easy access is provided by a lift inside the pit frame, thus linking the mine with the platform on top of the winding gear from where the visitors can enjoy a marvellous view over the central Ruhr, one of the most interesting landscapes on the Continent. There they can get an idea of how this industrial area has been developing, how the cities and traffic lines have grown, and how industrial sites have been built — and sometimes been abandoned.

THE DEMONSTRATION MINE

Without doubt the demonstration mine is the greatest attraction of the German Mining Museum. It has been constructed in the marl bed some 30 m above the seam-bearing strata, and is continuously being extended by the active (not retired) miners employed at the museum. That is why neither the coal nor the iron ore, which covers the latest extension, is actually mined; although the machines are set into motion by the guides, their operation stops some centimetres in front of the coal. Nevertheless, work in this pit is simulated under conditions which are as true to reality as possible, including the dimensions of the roadways and many of the characteristic details of the surroundings. Most of the visitors go underground to get an

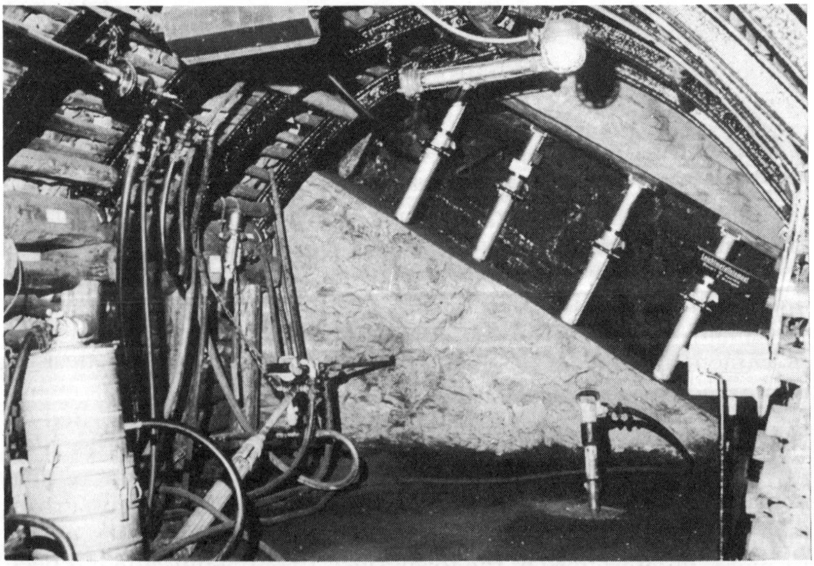

Figure 2 A flat seam in the visitors' mine.

impression of the usually hidden world of the miners, to sample the atmosphere of galleries, adits and coal seams, to experience the noise of the machines — in short the miners' working places — and to enjoy the feeling of being 'down below'.

The main purpose of the demonstration mine is to mirror the present technological standards of mining, to give an impression of how the miners' work place is governed by electronic equipment and remote control, by automatic processes as well as sophisticated safety measures. But what is more important, and makes it more than a mere fairground exhibit, is the fact that some of its older units and installations now no longer correspond to the latest technology. During the past four or five decades mining engineering has made decisive progress in productivity, and the combination of both these aspects yields the possibility of realizing rare didactic opportunities. On the one hand, it conveys impressions which one cannot gain when visiting a mine actually in production because of the multiplicity of experiences one is exposed to in real working conditions, and, on the other hand, the visitors can be instructed about outstanding developments in the sense of the recent history of technology.

One fundamental subject, for instance, is the fact that the visitors are moving in caverns which have been artificially constructed just for that purpose. Consequently, first of all, they should be informed how this has been realized. Two different types of machinery have been installed in order to explain the change that has taken place with this process. The first one is a compressed-air drilling carriage from 1956 showing how at that time single blast holes had to be drilled into the rock in order to be filled with explosives. When the firing crew had ignited the charges and the damp had vanished, the shattered rocks had to be loaded and conveyed, and the new section had to be secured by setting supports. It was a rather short-lived sequence, however, and, very quickly, all the single steps of the procedure had to be repeated. Manpower was needed in large quantities, the work was extremely tough, the machinery extremely loud, and drilling the holes was hazardous to health as the airborne dust often caused silicosis. Next to that drilling carriage, a tunnelling machine has been installed. Eighteen metres long and weighing 80 tonnes, it was installed in the museum's mine shortly after it had been used to build a water supply tunnel in southern Germany in the early 1970s. The machine was only a prototype, and more recent ones, admittedly, are even more gigantic and effective, but it represented an increase in productivity that was believed to be impossible before. In just a single operational sequence it cuts roadways with a diameter of several metres and conveys the cut rock continuously. One single worker operates the electro-hydraulic machine, a laser beam giving the direction of steering for hundreds of metres. This means relatively easy working, of course, but by comparing both units, the visitor simultaneously gets an idea of how modern technology is liable to make workforces obsolete.

Figure 3 Demonstrating the tunnelling machine in the visitors' mine.

Modern highly productive technology can also be studied by reference to another complex of machinery which is situated in the central part of the mine and which deals with the fundamentals of the miners' work — the getting of coal itself. This subject is tackled by a pneumatic pick in a small inclined seam and, by contrast, by a coal plough and a shearer-loader, both working in completely furbished flat seams with faces each more than 50 m long. In the 1920s the pneumatic pick became dominant as a new and highly welcome tool for extracting the coal from the seam. It replaced the heavy and tiresome work with the manually wielded pick that had previously been the only option. With the pneumatic pick, coal could be won mechanically, with compressed-air-driven machine tools, instead of relying strictly on the muscles of the miners; this brought new and higher levels of productivity. On the other hand, the pneumatic pick often caused diseases of miners' arms. The displays allow the visitors to imagine how a miner felt using such a drill during an eight-hour shift, lying on his back, in a seam sometimes no thicker than 0.5 m. In the 1950s, about twenty miners would have worked to get the coal from a face about 120 long; they would have been able to produce about 350 tonnes in one shift. An additional five miners were needed for loading the coal, and twenty-five more were required to secure the roof by drawing the wooden props and setting up new ones. Nowadays, on a modern coal face with a length of 200 m, more than 1,500 tonnes is produced during each shift. The coal-plough and the coal-cutter are doing the work, in many cases directed by remote control. Only ten miners work in such a seam, with roof supports consisting of variable-length steel props driven hydraulically.

THE GALLERIES

As the museum is situated in the centre of the Ruhr district it is often assumed to be concerned only with coal-mining. Even its new name, which was chosen carefully to give the impression of its more general nature, often does not prevent this misunderstanding. But as a matter of fact, a broader conception of the museum's purpose led to the acquisition of historically important objects from other regions and other branches of mining activity also. Gradually, this led to the presentation of the entire history of mining throughout the ages and even the continents. The museum's subject could not remain restricted to coal-mining only, as it is a relatively recent activity compared with for instance, the winning of iron ore, gold or the still earlier flint mining.

With all its activities — generally in accordance with the fundamental functions of every museum, e.g. collecting, preserving, researching and displaying — the museum seeks to reveal the complex relationships which exist among the various mining technologies, and it does not aim to present only isolated specific examples of engineering. Further, it makes an effort to set the technological aspects in the framework of their social and economic settings. It is moving away from simple descriptions of the objects: the idea is to place more emphasis on questions which consider the background, the foundations and the historical context in a more general manner. With this in mind, the museum, for instance, is devoting a spacious area to the subject of 'Mining and Environment'. Here is asked the question: in which ways and to what extent is mining directly or indirectly a part of the texture of human life, including its impact on ecology? Also considered are questions of surface damage and the changing of the landscape by waste-heaps, as well as the reclamation of open-cast mining sites, the growth of new towns, the agglomeration of townships and of entire industrial regions. The results of new public regulations and prescriptions are shown and contrasted with former practices.

At present, there are altogether some twenty-five different departments in the museum. Most of them, especially the older ones which are following their original purpose, are dealing with one particular branch of mining technology, starting with the very earliest tools and equipment and in most cases pursuing further development to the present.

One such example may be taken from the problem of shaft-hoisting, transporting men, goods and minerals into and out of a mine. The first machinery used in prehistoric times was a simple whim (a kind of windlass) on to which a rope was rolled and from which it was unrolled. Winding in the rope was as hard and monotonous as working the treadmill in later times. These conditions lasted until the Middle Ages, when water power came into use as a new form of energy that eased the work. But water-wheels could only be used where a certain amount of water was available. Lakes in the Harz Mountains, for instance, are today regarded merely as pleasant features of the landscape; it is generally forgotten that they were brought into being by man at high cost in order to supply water for the hydraulic machine systems in the mines, foundries and smelters throughout the seasons. Where no water was available it was necessary to develop other technical devices, such as the horse-driven gins which came into use at the end of the Middle Ages. Gins dominated the appearance of many mining regions in Europe before the invention of the steam-engine. Today's standard of technology is represented by winding-towers with the engine on the top. In this section, models up to several metres high demonstrate the most important achievements in hoisting.

Many of the models shown can be set into motion by pushing buttons, thus making mechanical functions easier to understand as well as being more attractive to visitors. This is also the case with a series of models that are on display in the Department of the History of Mine Drainage. As natural draining by using inclined drifts does not work when the mines are situated in strata below the bottom of a valley, since antiquity many a technical device had to be invented before the days of the steam-engine; nowadays fully automatic electric pumping units have become common. Thomas Newcomen's engine, among others, represents a major type in use over the centuries. Until then, when water to be used for draining was not at hand close to a mine (and this is also shown in this section), flat rods had to be constructed to transmit the source of energy. A large diorama gives an impression of the enormous technical and financial problems that resulted from the adoption of this technology.

Manpower, again, was the easiest and cheapest way to apply to mechanical devices in the earliest times. This is demonstrated, for instance, by a model showing large wheels to raise water in Roman silver mines.[6] Slaves did the work, and the lash of the foreman gives an impression of the working conditions in those mines. Notwithstanding that, the Romans may possibly have used piston pumps for the draining of mines — a fact that remained unknown to historians of technology for a long time. One of these pumps[7] was found in a copper mine in the Spanish province of Huelva in 1889 and it has been preserved in Madrid.[8] Made from twenty-six bronze parts in the first century AD, the force pump most probably had been installed in a trough from which the water was to be pumped. It consists of two cylinders with pistons and valves, and the area of flow is narrowed so much that the velocity of the water is increased twenty-five times; almost one and a half pints (0.8 litre) was raised by one stroke of the piston — too little perhaps for serious pumping purposes. Soon after the German Mining Museum became aware of this piston pump, its archaeologist contacted the Spanish authorities, following which engineers and craftsmen of the museum prepared the drawings in order to rebuild the machine at their workshops. It is only recently that the replica of this 'pearl among Roman antique finds' has become a paramount constituent of the museum's collections.

More recently, the skills of the museum's model builders and mechanics have been devoted to another project related to pumping: restoring one of Georg von Reichenbach's water column engines (its pistons being driven by a water column, the water thrusting down by a beam from a special reservoir). Originally, in 1809, the engine had been installed at a pumping station lifting brine into a reservoir 97 m higher, and it was a part of the famous line that transported brine from Reichenhall to the salt-works at Rosenheim in Bavaria[9]. The engine was double-acting, like Richard Trevithick's machines; this second type of Reichenbach engine had only one pumping cylinder, being connected with the two pistons by a plunger rod. All its parts were of metal, were easy to make and proved to be extremely durable; indeed, the engine had been working by day and by night for almost one and a half centuries until the line and the salt-works were abandoned in 1958 and the machine was transferred to Bochum. After long-standing difficulties experienced throughout the years in housing such a large object within the museum's collections in a setting appropriate to its outstanding character as an industrial monument, plans have been settled to place it in a centrally located outside area, protected by a specially built glass house. Former plans to have it driven again by water power have been abandoned because of the high costs of maintenance. Instead of that, strictly to Reichenbach's original drawings,[10] a 1:7 scale model is under construction which will be displayed together with the historic engine. As it will be set into motion by water power it must be built completely of brass.

The practice of combining original historic objects with models is one which the museum learned in its early days. It seems to be a very useful way of transferring knowledge to visitors, especially where

Figure 4 Original, partly reconstructed steam-engine of Königsborn salt-works from 1799, the first in the Ruhr area.

technological items and, even more, specific details or full explanations of their engineering are concerned. So, this didactic principle will be continued despite the high labour costs of maintaining specialized workshops under current conditions.[11] To give another example, it also helps the visitors to understand the rather complicated aspects that are involved in the technology of deep drilling. In this collection of the museum several diagrams give an overview of methods in use through the ages, beginning with the Chinese method

Figure 5 Modern loading equipment in metal-mines.

of rope drilling, then describing machines used in the eighteenth century, and ending with a synopsis of the specific situation in the nineteenth century, with the inventions of Jean-Baptiste Jobard in Belgium, Karl von Oeynhausen in Germany, and Edwin Laurentine Drake in Pennsylvania. The technique of rotary drilling has become common since then. The method of Joseph Chaudron and Karl Gotthelf Kind for sinking shafts of full diameter is represented by a special model which can be set into motion. Another rather large-scale one shows Anton Raky's rapid boring rig of 1895 by which deep holes could be sunk at speeds hitherto impossible.[12] His drilling method was unrivalled and became predominant on the Continent until improved rotary drilling rendered it obsolete. A large variety of the tools used with the different methods are on display in this collection. They are replicas rather than the originals, although original equipment is preserved in the museum's collections also.

But the German Mining Museum is not solely concerned with presenting the history of technological developments, and not all of its galleries are concerned with such aspects only. The collection dealing with the genesis of mineral deposits and the economic value of raw materials, including many graphic layouts, exemplifies this, as does the collection which illustrates the influence of the mining industry on fine arts and culture as a whole. The history of civilization, reflecting the interplay of distinct political, economical, social and artistic influences, finds expression, *inter alia*, in works of art relating to mining subjects. Mining activities influenced the fine arts and cultural life in general, as well as influencing social history and the history of technology. In order to show this as comprehensively as possible, the museum includes in its collections replicas or photographic reproductions of relevant works of art owned by other museums. In this case the documentary value is deemed more important than mere illustration. That is the reason why, in addition to an enlargement of the front page of the *Kuttenberger Kanzional* (a Bohemian collection of spiritual songs from the first decades of the sixteenth century) being included, there are also some twenty slides showing details of many items of technical interest, this linking up with subjects addressed in other collections of the museum. As far as social history is concerned, this rare iconographic source, as well as the *Annaberger Bergaltur* in Saxony from about the same time, conveys important information on the role of women in early mining. 'Women at work' is the theme of a specific permanent exhibition at the museum, here dealing with a particular chapter in the history of mining long neglected by researchers: The themes will be pursued within a joint programme with museums overseas.

Another example which illustrates the broad thematic spread of the collections concerns silver-ore mining in the Middle Ages. This is the result of more than a decade's research work on a mining district in southern Westphalia that is more usually 'celebrated' for its iron ore. Scientists from the museum have carried out surveys and excavations on the mountain site and in a former miners' camp which, legend says, had been destroyed as a sign of God's justice since the miners had become too rich and had strayed from the common rules of morality. One of the manifestations of this, according to the tale, was that they started gambling and playing bowls instead of doing their work. The archaeologists, in fact, not only discovered several shafts and ancient miners' dwellings, but even found the relics of the oldest bowling game ever known. Mining operations on this site could be dated to the early thirteenth century. Wooden shaft supports were also found, so the technology of shaft sinking could be verified for a period for which there was no literary evidence. The manifold finds made there laid the foundation for an attractive gallery featuring a reconstructed shaft, incorporating some of the original timber supports. Many other exhibits and reproductions, and a lively graphic design, recreate the atmosphere of mining life almost seven hundred years ago, during an important period in the history of mining that had hitherto remained untackled by authors and by museologists because of the lack of real evidence.

INTERDISCIPLINARY RESEARCH WORK

Admittedly, the vivid character, as well as the information which gives a distinct atmosphere to this section, cannot easily be realized in every gallery. It may be regarded as the ideal approach but it goes without saying that the realization of any such scheme resides above everything else in the quality of historical imagination that is brought to bear on the work in hand. Ideally, each reconstruction should involve close co-operation between the craftsman, the architect, the designer and the research staff — a collaborative effort that initiated the reconstruction of the Roman bronze pump and the completion of the Reichenbach engine with its additional display.

Interdisciplinary research work involving engineers, archaeologists, historians and scientists has initiated the present scheme of supra-regional and national budgeting of the museum. It has turned out to be a solid basis for the aims the German Mining Museum is pursuing. Its results, also in respect in technological developments, are mirrored in many publications, especially in the museum's own book series[14] and regularly in the bi-monthly international review on the history of mining, *Der Anschnitt*, which is also edited by museum staff and published by its sponsoring association.[15] This general concept has proved to be the more important when minor, rather than major, inventions are to be studied. Without such a basic perception of staff work many important items of the history of mining could hardly be researched comprehensively, and its necessity becomes the more evident when, for instance, lesser or even misleading developments have to be evaluated.

For research subjects like these, the museum is in the favourable position that in 1969 the Mining Archives had been established as a special department of the museum and as a national documentation centre of written source material. This foundation was the museum's reaction to the reorganization of the Ruhr's coal-mining industry and the subsequent shutting down of collieries. At that time the question arose as to where and how to keep the files which individually and *in toto* are regarded as most important for the historiography of mining. The Mining Archives Department takes over stocks deriving from former mining companies, associations or private individuals. After having each item calendared and, if necessary, preserved, surveys of such source material for further research work are presented in the form of reference books.[16] The demand in the learned world of universities for such information becomes constantly more evident. In the present situation, and for the future of the mining industry in Germany, the importance of mining archives is still growing.[17] Besides written documents, the many drawings and plans originating in the technical departments of the companies are of specific evidence when technological developments are in question. The difficulties arising from the not only sceptical but almost hostile attitude of his company that Friedrich Koepe was confronted with as he endeavoured to prove the worth of his later world-famous idea of the friction-winder for shaft hoisting in practice[18] may be mentioned as just one example of the importance of preserving for historians company papers that throw light on just such situations as these.

Notes and References

1. On this subject see the recent summary, Werner Kroker, 'Das industrielle Erbe des Bergbaus und seine Pflege im Ruhrgebiet', *Acta Archaeologica Lovaniensia*, 1992, 31: 49–58.

2. Werner Kroker, 'Bergbaugeschichtliche Stätten im Muttental bei Witten', *Der Anschnitt*, 1974, 26: 30–7.

3. Rainer Slotta, *Technische Denkmäler in der Bundesrepublik Deutschland* (Bochum, 1975). From this date several volumes, each on special branches of industry, by the same author.

4. Werner Kroker, ed., *SICCIM: Second International Congress on the Conservation of Industrial Monuments. Transactions* (Bochum, 1978); since the conference in Bochum in 1975 there have been several TICCIH conferences in Europe and in North America.

5. Evelyn Kroker, co. ed., *50 Jahre Deutsches Bergbau-Museum Bochum; Fotodokumentation* (Bochum, 1981), 38–84; see also *Deutsches Bergbau-Museum Bochum/German Mining Museum*, English edn (Braunschweig, 1979).

6. Covered very extensively in Franz Kiechle, *Sklavenarbeit und technischer Fortschritt im römischen Reich* (Wiesbaden, 1969).

7. Thorkild Schioler, 'Bronze Roman Piston Pumps', *History of Technology*, 1980, 5: 17–38, esp. 20–8.

8. At the Museo Arqueológico Nacional.

9. For the best technical description of the engine see Walther v. Dyck, *Georg von Reichenbach* (Munich, 1912), 61–8.

10. They have been preserved in the plan collection of the Deutsches Museum in Munich.

11. For a more detailed description of the concept, which is still valid, see Werner Kroker, 'Urproduktion und moderne Technologie: zur didaktischen Vermittlung des Bergbaus und seiner Geschichte', *Geschichte lernen im Museum*, eds Annette Kuhn and Gerhard Schneider (Düsseldorf, 1978), 138–50.

12. For an evaluation of the technologies, see Dietrich Hoffmann, *150 Jahre Tiefbohrungen in Deutschland* (Munich/Hamburg, 1959).

13. See the catalogue *Frauen und Bergbau: Zeugnisse aus fünf Jahrhunderten* (Bochum, 1989).

14. Veröffentlilchungen aus dem Deutschen Bergbau-Museum, 53 vols (1969–94).

15. *Der Anschnitt*, in its 46th year, is published by *Vereinigung der Freunde von Kunst und Kultur im Bergbau e.V.*: for the contents of vols 1–35 (1949–83) see Werner Kroker and Evelyn Kroker, eds, *Der Anschnitt. Generalregister* (Bochum, 1985).

16. For a complete listing, see Evelyn Kroker, *Das Bergbau-Archiv Bochum. Kurzführer*, 2nd edn (Bochum, 1994).

17. For the role and functions of these institutions see Evelyn Kroker, 'Business Archives in the Federal Republic of Germany', *Business Archives*, 1993, 65: 51–60.

18. Hartmut Arnold and Werner Kroker, '100 Jahre Schachtförderung nach dem System Koepe', *Der Anschnitt*, 1977, 29: 235–41.

The Structure of Technological Revolution

R. ANGUS BUCHANAN*

Social scientists and historians of science will not need to be reminded that this title is designed to echo that of the American historian Thomas Kuhn, whose seminal treatise *The Structure of Scientific Revolutions* was published in 1962 and promoted the term 'paradigm' into our professional vocabulary. Kuhn developed the intriguing thesis that modern science, far from developing as a uniform progressive movement from simple origins in the sixteenth century to its present complexity, has actually advanced in a series of conceptual jumps which he called 'scientific revolutions'. Each of these was characterized by a sudden change of basic assumptions which followed a prolonged period of 'normal science', distinguished by general agreement concerning such assumptions, and followed by another period of relatively little change during which the implications of the new body of shared assumptions was worked out in detail. Kuhn called these bodies of assumptions 'paradigms', and the scientific revolutions by which science jumped from one set of assumptions to its successor thus became 'paradigm shifts'.[1]

Kuhn was able to produce a convincing number of occasions on which such paradigm shifts had occurred, and even though aspects of his argument have been subjected to damaging philosophical and historical criticism, most scholars in this field continue to find his thesis helpful. Useful as it may be to historians of science, however, the Kuhn thesis does not comfortably fit the history of technology. The main reason for this is that, whereas science is primarily a conceptual exercise concerned with our understanding of ourselves and our environment, technology has always been concerned mainly with more matter-of-fact issues of physical power, productivity, and the ability to make and do things. Strictly speaking, even though the word is almost invariably used loosely in English, "technology" is concerned with the study of these things rather than with mastering the actual techniques of making and doing, so that it does possess an important dimension of conceptual understanding. But the emphasis is on physical artefacts and achievements rather than ideas: as Lord Ashby once said, in a memorable phrase, technology is 'of the earth, earthy'.[2] It follows from this that technology is not responsive to intellectual paradigm shifts in the same way as science. So while historians of technology should be grateful to Kuhn for alerting them to the importance of the structure and internal coherence of their discipline and, incidentally, for demonstrating the inadequacy of the simplistic progressivism which has passed for scholarly assessment in too much of the 'wonderful inventions' sort of history of technology, they need to look elsewhere for a satisfactory structure of technological revolution.

Attentive readers will have noticed that I use 'technological revolution' in the singular, in contrast with Kuhn's plural use of 'scientific revolutions'. My reason for doing this follows from what I have already said about the 'earthiness' of technology. We are not dealing here with a mass of individual perceptions and conceptualizations which are capable of sudden changes of direction, but with physical hardware and actual productivity which, despite widespread variety of form and intensity, are capable of treatment as a single cumulative movement and possess a certain common momentum which makes them resistant to sudden change. That is not to say that the process is necessarily uni-directional: the history of technology has recorded several cases of technological break-down when the technological achievements of civilizations have been lost or destroyed. But in referring to 'technological revolution' in the singular, I am merely observing that the sensational technological achievements of modern history, and especially those associated with Western civilization, form part of a remarkably uniform process of development which has continued now for at least three centuries without sharp discontinuities or reversals. The problem for the historian of technology is to understand how and why this process has happened, and that is why it is important to penetrate into its inner mechanism and to determine its structure.

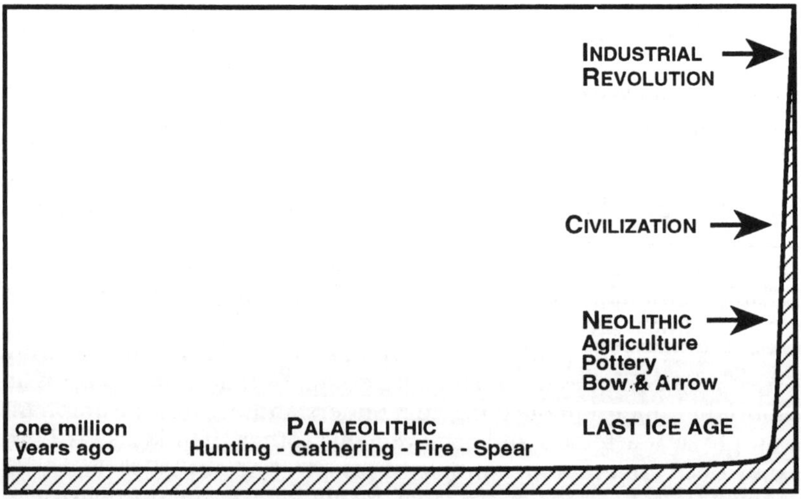

Figure 1 Technological development over 1 million years.

Before we embark on this enquiry, however, it will be prudent to enter a few caveats in order to limit the scope of what could otherwise be an almost boundless exercise. In the first place, while restricting our attention largely to Western civilization, it is important to recognize that the process of technological revolution has much longer roots, going back at least ten thousand years to the period of the so-called 'Neolithic Revolution' (Figure 1) when people of the New Stone Age acquired techniques of settled agriculture and the domestication of animals which produced a marked increase in productivity and which led in turn to a population explosion and the beginning of urban civilization. The technological revolution with which we are concerned thus represents an acceleration in a long-standing process of technological development and social change. Secondly, I would like to stress that, in the simplified account which is all I can offer here, I am not trying to revert to a monocausal explanation of social change equivalent to those now discredited historical interpretations which once sought to explain everything in terms of national characteristics, climatic factors, or even class war. Rather, I am trying to emphasize an aspect of a complex process which has frequently received less attention than it deserves, but I would not deny the continued importance in social change of political, religious and cultural factors of which I will have little to say. And for my third caveat, I would like to observe that, while much of my treatment will be concerned with those inventions and processes which have been manifestly successful in changing society, I am not unaware of the interest and possible significance of the many failures in the history of technology. Only I would also observe that the subject is inevitably asymmetric, for the simple but incontrovertible reason that it is the successful inventions which have (by definition) produced a marked social impact, and it is with the consequent social transformation that technological revolution is concerned.[3]

The association of technological revolution with Western civilization has been one of the most significant facts of world history: indeed, it could be said tht it is unique to Western civilization for, despite some outstanding technological achievements in other cultures, it is only in the West that it has established continuity of development over several centuries. The early centuries of Chinese civilization provide an interesting precedent, as there is no doubt that China led the world in technological development until the end of the first Christian millennium (Figure 2), with astonishing innovations in metalworking, ceramics and navigational techniques long before any of these had been introduced in the West. Thanks to thc monumental research of Dr Joseph Needham, it is now beyond any reasonable doubt that crucial innovations such as windmills, gunpowder, paper, printing, the manufacture of brass, the mechanical clock and the production of porcelain were all passed on to Western civilization from China. But whereas these and other innovations triggered a chain of social revolutions in the West, their influence in China was

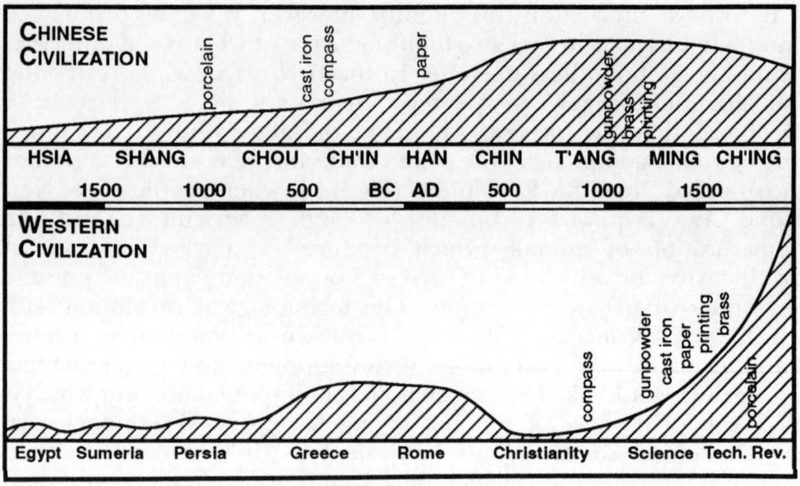

Figure 2 The comparative development of world civilizations. The vertical axis represents stability, complexity, urban development, etc.

controlled by the professional bureaucracy of the mandarins in the interests of an efficient centralized government, and was never allowed to become socially disruptive.[4]

The example of gunpowder is instructive, for even though Needham has demonstrated that it is not accurate to regard the Chinese as having produced it only for demonstrative purposes in fireworks and other luxuries, it failed to have the revolutionary effect on warfare which it caused in the West — at least, until it was reintroduced into China by Western gunboats in the nineteenth century. Whichever way one looks at the long centuries of Chinese civilization, the role of technological innovation has always been controlled and functional rather than revolutionary. And other great world cultures, such as those of the Indian and Arab civilizations, showed even less interest in the applications of technological innovations until they, also, were taught to do so by the West. The role of the West, in this respect, has thus been extraordinary, and it is a matter of tremendous current importance that the rest of the world is now looking to the West for leadership in technological revolution. When I was invited to lecture to a group of Chinese teachers in 1983, at the Central China University of Technology at Wuhan on the Yangtze, it soon became apparent to me that what my audience most wanted to learn from me was the secret of Western technological revolution whereby, rightly or wrongly, it seems to the perceptions of many intelligent people in the developing world that the source of Western wealth, power and productivity can be discovered.

So what is the nature of the process of technological revolution by which the West has aquired the ability, quite literally, to produce many ears of corn where only one grew previously, and to make available unprecedented sources of physical power and material prosperity? Obviously no answer I am able to give in a few words to this question, about what is an enormously complex process, is going to be adequate, but I would like to make a couple of suggestions about the lines on which an answer might be framed. One of these is concerned with the mechanism by which the process works, and the other with the social construction which makes it possible. First, then, it is important to observe the cumulative nature of innovations. Note that I am talking about innovations – successfully applied inventions – rather than inventions themselves. Identifying the source of inventions is an even trickier problem, involving the psychology of human inspiration and creativity, and I will leave that on one side for the moment except to say that it is reasonable to assume that any human society will produce inventors: it is the use that society makes of their ideas which is significant for our present purposes. And in so far as the successful application and development of innovations has been an outstanding feature of modern Western societies, the cumulative quality of the process has been integral to its success. Innovations have always drawn on existing techniques in their early stages, and have then frequently provided the necessary starting skills for the techniques which have superseded them.

This 'ratchet' process, by which the development of every innovation follows a form of 'S' curve in which its period of maximum expansion is both preceded and succeeded by a period of slower growth, emphasizes the way in which every technological advance draws from previous experience or helps to form following innovations. Thus the canal builders of the eighteenth and early nineteenth centuries learnt much from the fairly simple road-building and estate-surveying techniques of their predecessors, and passed on more sophisticated techniques of tunnelling and embanking to the railway engineers who came after them; the development of the bicycle depended heavily on traditional skills of wheel-making, but within a few years passed on tangentially spoked wire wheels with pneumatic tyres, chain drives and ball bearings to the new automobiles; and twentieth-century radios drew on knowledge of telephony and soon passed on features of structure and style to the original television sets. The connections between the existing successful technologies and the new ones may be of various types. They may be simply inspirational or suggestive, in the sense that they provide the confidence necessary to undertake a new development. Or they may supply actual hardware: iron cylinders, pistons, ball-bearings, and suchlike. Or again, they may stimulate novelty by their inadequacies, as when James Watt applied himself to discover why a model Newcomen engine did not perform as efficiently as expected. In whatever way it occurs, however, the 'ratchet' relationship is a vital incentive to the formulation and successful launch of new technological ideas.

One of the most important examples of this 'ratchet' process is

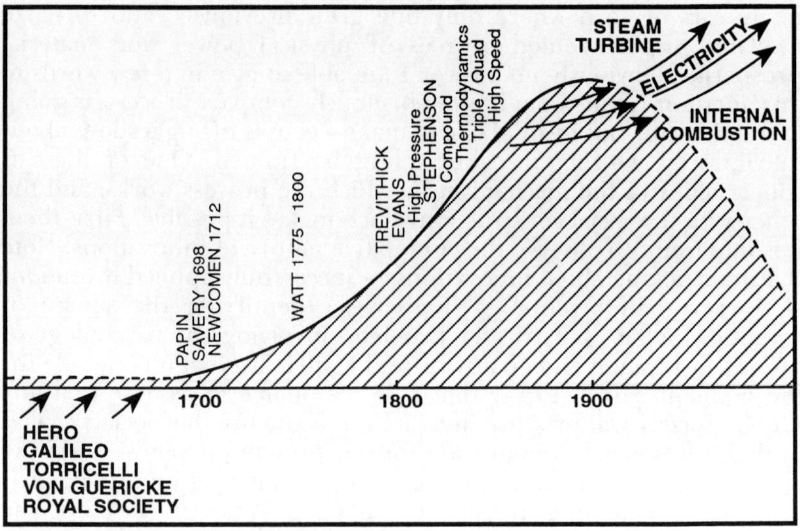

Figure 3 The development of reciprocating steam power, demonstrating the 'ratchet mechanism' of a successful innovation.

provided by the history of steam technology (Figure 3). It is not only that, having become almost completely obsolete, the reciprocating steam engine provides a wonderfully complete instance from recent history of the origin, growth and decline of a most successful innovation, although it does all that. My choice is also justified by the fact that the Centre for the History of Technology acquired considerable expertise in the history of this subject through the work of my late research colleague George Watkins. The superb Watkins Collection of photographs of stationary steam engines is now held by the Royal Commission on Historical Monuments in England, and George prepared five selections of the best of these for publication, annotated with characteristic insight. His work has brilliantly illuminated the final stages in the development of stationary steam technology.[5]

The first successful steam engine, creating a vacuum in a cylinder by the condensation of steam, and using the consequent atmospheric pressure to produce the downward stroke of a piston in the cylinder, was installed by Thomas Newcomen at a coal-mine near Dudley in 1712. The immediate processes by which Newcomen arrived at this outstandingly novel invention are obscure, because we know so little about the man, but it is clear that it engaged with existing knowledge at the end of the seventeenth century about the weight of the atmosphere, the nature of a vacuum, and the possibility of creating such a vacuum with an air pump. It also drew on existing knowledge of water pumps, with a plunger operating in a barrel, and of Denis Papin's steam pressure vessel and Thomas Savery's steam pumping contrivance, as well as on many generations of water-pumping and power transmission technology. Once it had established itself as a successful technique, the Newcomen engine became immensely popular as a pumping machine in the coalfields of Britain, where the heavy fuel consumption of the engine was not a serious problem. The ratchet process continued to operate as subsequent inventors worked to improve the performance of the steam engine: James Watt introduced the separate condenser, which brought substantial savings in energy, and followed this with double action (the piston producing power on the up-stroke as well as on the down-stroke), rotative action (the simple reciprocating motion of the piston in the cylinder being converted to rotary movement suitable for driving mills and wheels) and other refinements; Richard Trevithick and fellow-Cornishmen developed the high-pressure engine; and a series of ingenious engineers in the nineteenth century brought in compound engines, triple-expansion engines and high-speed engines in a wide variety of shapes and sizes. At every stage they were able to make considerable improvements within the existing technology of steam power, building always on the skills and achievements of their predecessors.

In these circumstances of a highly successful technology it is difficult to recognize the moment of 'technological obsolescence', when the capacity for further improvement falls behind the potentiality of replacement by a different technology. Such a moment is discernible clearly enough in retrospect when the old technology has been supplanted, but until that has happened the possibility of improvement remains valid. Consider, for instance, the recent history of the automobile. Two decades ago, the possibility that the conventional recipcrocating engine would be replaced by the rotary 'Wankel' engine seemed on the point of becoming a probability, but in the event the conventional engine was improved sufficiently to head off the challenge, thus delaying the moment of technological obsolescence and consigning the Wankel engine prematurely to the technological museums (even though it is now showing renewed promise of development for specialized purposes).[6] In the case of the reciprocating steam engine, the arrival of alternative technologies did not immediately seal its fate. Charles Parsons developed the steam turbine out of the existing technology, and the early years of the internal combustion engine as invented and improved by Lenoir, Otto, Benz, Daimler and Diesel were similarly conditioned by the techniques and experiences of steam technology. Only gradually did the new technologies displace the reciprocating steam engine from electricity generation, from large industrial applications, from marine transport and, finally, from rail transport. Traditional steam technology could not compete with the newcomers in economy, thermal efficiency or convenience of operation, but in every case it provided the starting ratchet which set the new technologies in motion.

These examples serve to establish a typical mode by which technology advances. But, needless to say, they do not determine a pattern which must always be followed, and it is certain that some new technologies derive from inventions which were relatively unexpected: Edison's phonogaph, powered flight, nylon and the transistor could all be placed in this category. Even with inventions such as these, however, it is possible to assemble an impressive record of previous expertise which became relevant to the needs of the moment when the innovation was introduced, so that a form of the ratchet process is still operative. What is more intriguing, from the point of view of the structure of technological revolution, are the reasons for the operation of the ratchet which we have been describing. After all, it is not necessary that changes to more powerful, more efficient or more convenient techniques should always be adopted, and there are strong social factors of technological momentum and cultural conservatism which regularly act as constraints on such changes. It may be argued that it was an over-commitment to steam technology which signalled the decline of the British entrepreneurial classes at the end of the nineteenth century, making them less receptive than their competitors to the new technologies of electricity and internal combustion, although it is difficult to maintain such an argument with an adequate basis of quantitative data. Whatever the pros and cons of this particular argument, however, it serves to make the point that reactions to technological innovation are socially determined, and it is to this aspect of the modern technological revolution that we must now turn.

At a very superficial level, nobody would deny that there is an element of social construction about technology. The adage that 'necessity is the mother of invention' has, after all, passed into the mainstream of Western consciousness, and the evidence of response to social need is readily to hand: the Newcomen steam engine, for instance, was a response to the need of the mining industry for better pumping equipment (although Newcomen was thinking more of the needs of the tin and copper mines of Devon and Cornwall than the coal-mines where it was actually adopted), and continuing needs for ever greater productivity and facility to transport goods and people have promoted many of the most dramatic inventions of modern times. But, equally clearly, the existence of a social need is not itself enough to generate innovation. The social needs of an Eastern peasant labouring in his paddy fields are at least as great as those of an urban worker in a Western city, but there has been no flow of innovative equipment to lighten his labour and enrich his leisure commensurate with that which is continuously transforming the conditions of life in our Western cities.

So needs require the stimuli of other factors before the ratchet of improvement can begin to engage. Social resources, for instance, are a prerequisite to any successful innovation, and among these the most important are capital, materials and technical expertise. Shortage of capital inhibits much worthwhile technological improvement in developing countries, preventing the installation of electricity, piped water and expensive mechanical equipment. Just how worthwhile these things are is admittedly a matter for debate, now that we have become conscious of the arguments for small being beautiful, and for intermediate or alternative technologies.[7] But the fact which concerns us here is simply that they are not available without the resource of wealth or capital to invest in their procurement, and the lack of this particular social resource has been a very significant limitation upon the scope of technological revolution.

Similarly, lack of suitable materials has prevented the successful fabrication of many promising inventions. Leonardo da Vinci's drawings show brilliant ideas for submarines, aeroplanes and helicopters which could not be made at a time when structures were still built mainly of wood and stone, and iron technology was fairly primitive. Even in the twentieth century such limitations have affected the course of innovation, as when the ideas of Frank Whittle for a gas turbine engine were delayed for a decade until alloys became available to resist the high temperatures of the combustion chamber; and the problem of bringing a space capsule back to earth without it burning up in the atmosphere seemed insuperable until new ceramic materials were developed to provide a heat-shield.

Again, lack of the resource of technical expertise has been a potent force in preventing the adoption of inventions. James Watt's separate condenser required a high standard of accuracy in machining large pieces of metal in order to secure the advantages which made it successful, and he had great difficulty in finding craftsmen in Scotland who could perform such work. It was only his partnership with Matthew Boulton, the Birmingham businessman who possessed a large labour force of skilled metalworkers, which brought him into contact with the technical expertise he needed. Only then, in 1775, six years after taking out his original patent, could the manufacture of his improved steam engine be undertaken in earnest. The same need for suitably skilled craftsmen has become an important consideration in the growth of the modern electronics industry, and it has significant ramifications for the role of technical training, which modern societies neglect at their peril.

But even the addition of these social resources to the factor of social need does not guarantee a successful process of innovation: there still needs to be a source of ideas, and these in turn require a congenial social milieu if they are to flourish. We have already noted that the conception of technological inventions is a process of human creativity similar to that which produces fine art. It follows that it is a matter of individual genius for which no detailed plan can be preserved. On the other hand, it is possible to remove impediments to such genius, and to encourage it through various incentives. Likely candidates can be chosen and given the resources they need to think and to experiment, and any 'intellectual property' which they produce in the form of inventions can be protected by appropriate patent law. All modern industries work on the assumption that, by laying on 'research and development' teams and giving them proper encouragement, it is possible to generate a stream of inventions likely to be of value to the enterprise. Such hopes have been amply vindicated by many developments in electrical engineering, electronics, and chemical engineering, to name only a few of those which have benefited from such initiatives, and some of the most spectacular twentieth-century innovations — like the transistor, invented in the Bell Laboratories in the USA — have sprung from the work of highly co-ordinated research teams. While it remains true that all inventions — and the transistor is no exception — must in the last resort derive from the inspiration of some individual or individuals, and that most inventions are still produced by 'amateur' men and women of talent, there is no doubt that much can be done to 'shepherd' inventors into congenial research teams and thus to place their inspiration at the disposal of large organizations, including the state. The latter instance is particularly relevant in time of war, which has been notoriously effective in stimulating inventions, although it should also be said that any positive gains have usually been offset by the destructive effects of war.

Such a disposition to plan for successful innovation implies a further social precondition, which is perhaps more important than any of the others. This is the provision of an environment which does not merely tolerate inventors, or seek to spoon-feed them for the purposes of particular commercial or national enterprises, but welcomes them as representatives of a culture in which change is regarded as wholesome and beneficial. I will call this an 'open culture', by analogy with Popper's 'open society', and the similarity extends to a common commitment to democratic institutions.[8] Just as these maintain the integrity of individual freedom against the oppression of tyrannical political forces in Popper's model, so the 'open culture' is one in which democratic institutions provide the tolerance for novelty and individual variety which are anathema to 'closed cultures' of a monolithic or totalitarian society. Not that political democracy alone can guarantee such tolerance: a democracy can use ridicule and more powerful sanctions to produce conformity, as Socrates discovered. It is said of Sir George Cayley, the British aristocrat who did so much to generate the theory of aeronautics in the nineteenth century, that he was dissuaded from conducting experiments with flying machines by the ridicule of his neighbours. But at least democracy removes most of the political constraints on innovation, and opens up the possibility of inventors being rewarded for their genius.

To make the point in slightly different words, what I am arguing is that sustained successful innovation relies upon a 'packet' of social conditions which include a high degree of individual freedom, toleration for new and unfamiliar points of view, an enlightened programme of technological education, and a legal system of patent law which enables inventors to gain some financial reward for their acumen. When the Chinese students in Wuhan, whom I have already mentioned, showed their eagerness to learn the secret of technological revolution in the West, it was to such a packet that I felt obliged to refer them. It is true that much can be done by the direction of a dictatorial government — whether by a tyrannical individual or by a party — to copy 'best practice' technology and even to secure short-term advances. Adolf Hitler's programme of German rejuvenation in the 1930s and the successes of the Soviet space programme in the 1950s were both dramatic illustrations of this potentiality. But Hitler's irrational disruption of the Nazi weapons development programme showed the vulnerability of technological innovations in such a society, and the recent history of the 'command economies' of Eastern Europe has demonstrated their inability to sustain a posture of high technology. In short, technological revolution depends on an 'open culture' packet, and the fact that it has only occurred in Western civilization, and even there only in the countries which have approximated most closely to such cultural preconditions, is the most eloquent confirmation of this concept.

The structure of the process of technological revolution which has so drastically transformed life in the modern world thus possesses several distinct and definable qualities, even though the whole process defies simplistic attempts to pigeon-hole it according to the stereotypes of economic analysis or systems theory. Such theoretical constructs have their place in sorting out some of the initial confusions in understanding the nature of technological revolution, but they are unable to do justice to the scope and narrative complexity of the whole process. What can be said is that the process appears to advance by cumulative interconnections for which the image of the ratchet seems most appropriate, suggesting as it does the necessity for backward and forward linkages for every successful innovation. Furthermore, careful study of the process over recent centuries suggests that this ratchet process of cumulative advance operates most successfully in the social conditions associated with the packet of the 'open culture'.

I hope that it will be clear that I am not attempting any value judgement in my analysis of technological revolution: it is a matter of historical fact which requires recognition and understanding, but it is not necessary or desirable to regard it as an unqualified 'Good Thing'. Indeed, when the enormous damage inflicted by technological apparatus is considered — both in warfare and in terms of environmental degradation — it should be palpable that technological revolution has been the source of many serious problems, for most of which we have yet to find satisfactory solutions. The confident expectation of universal progress enjoyed by many nineteenth-century exponents of industrialization and of the benefits of technology have now disappeared, and I would not wish to resurrect them. Technology, after all, has provided a set of tools which humankind has used for good and ill, and what we must look to is the capacity of men and women to benefit from these tools rather than to be cursed by them.

Forty years ago, Herbert Butterfield wrote passionately and profoundly about the quality of judgement in history, by which moral imperatives operate to bring a nemesis to cupidity and arrogance.[9] He was writing of the mainstream of Judaeo-Christian morality, but as he was also virtually the first professional historian to take the rise of modern science and technology seriously, it is not unreasonable to infer he meant that our ultimate fate as a world civilization will be determined by our ability or otherwise to control the vast forces, technological and otherwise, at our disposal. Such a humane way of regarding the process of technological revolution should help to emphasize the importance of what Butterfield called 'man-in-history' rather than 'man-in-nature'. A generation later, now that computer microminiaturization, information technology, bio-engineering and the opening up of the frontiers of space research have added immensely to the god-like powers available to humankind, I see no reason to modify this view. We remain the creatures of our artefacts, and if we cannot learn to use them properly, they will surely destroy us.

Finally, however, it is possible to perceive grounds for hope in this rather sombre prognostication. The fact that technological revolution has brought humankind to the brink of self-destruction is intimately linked with its achievement in bringing us to the threshold of an incomprehensibly rich future in the exploration of the universe. The same rockets which cast a dire shadow over the closing stages of World War II also opened the way to the stars and, in the process, brought the world to some sort of glimmering recognition of the fact that, as we are all members of Spaceship Earth, we must live together or perish together. Thus technological revolution presents us with a challenge at the end of the twentieth century: the choice between continuing on the cheerless course of spoiling our own nest, or of finding a new unity which will enable us to undertake our cosmic mission.

Notes and References

* This essay was first delivered as an Inaugural Lecture at the University of Bath on 17 October 1991. The theme of the essay is developed more extensively in the author's *The Power of the Machine* (London, 1992).

1. Thomas S. Kuhn, *The Structure of Scientific Revolutions* (Chicago, 1962).

2. Eric Ashby (Lord Ashby), *Technology and the Academics* (London, 1959); the reference to 'earthiness' is on p. 66.

3. For a treatment of the theme of 'Failed Innovations', see the ICOHTEC Symposium reproduced in *Social Studies of Science*, Vol. 22, No. 2, May 1992.

4. Dr Joseph Needham has devoted much of his long life to the monumental study, *Science and Civilization in China*, the first volume of which was published by Cambridge University Press in 1954. The treatment of gunpowder is in Vol. V, Part 7, published in 1986. Needham's general views about the influence of the mandarinate on technological innovation are more accessible in his *The Grand Titration* (London, 1976).

5. The first and most generally useful of the five volumes by George Watkins is *The Stationary Steam Engine* (Newton Abbot, 1968). For an overview of steam technology, see our joint work, R.A. Buchanan and G. Watkins, *The Industrial Archaeology of the Stationary Steam Engine* (London, 1976).

6. Developments of the Wankel rotating piston engine are now being used in some high-performance automobiles, so the story is by no means ended.

7. Studies of environmentalism, intermediate technology, etc. have abounded since the pioneering work of E.F. Schumacher, *Small Is Beautiful* (London, 1974).

8. K. Popper, *The Open Society and its Enemies* (London, 1945).

9. Herbert Butterfield's lecture series of 1948 at Cambridge University was published as *Christianity and History* (London, 1950). His study, *The Origins of Modern Science* (London, 1949), is certainly one of the earliest recognitions by an orthodox academic historian of the importance of the subject: 'it outshines everything since the rise of Christianity and reduces the Renaissance and Reformation to the rank of mere episodes, mere internal displacements, within the system of medieval Christendom' — Introduction, p. vii.

Reforming ICOHTEC: International Committee for the History of Technology

HANS-JOACHIM BRAUN

ICOHTEC REFORM

Founded in 1968, when the Cold War was being waged with particular bitterness between the nations of the Eastern and Western worlds, ICOHTEC provided a forum at which scholars from both sides could meet and communicate about matters of mutual interest in the history of technology.

ICOHTEC was constituted as a Scientific Section within IUHPS/DHS, the International Union for the History and Philosophy of Science (Division of the History of Science), at the XIIth International Congress for the History of Science, Technology and Medicine, held in Paris in August 1968. Through IUHPS it is associated with UNESCO, but it has always functioned as a fully autonomous body with its funding depending on its own members. Its first president was Professor E. Olszewski of Poland, with vice-presidents being Professor S. V. Shuchardin of the USSR and Professor M. Kranzberg of the USA. The first secretary-general was Professor M. Daumas of France, through whose initiative the French Government acted as hosts to the first independent (as distinct from those held in conjunction with the International Congresses) symposium of ICOHTEC, held at Pont-à-Mousson in July 1970. Symposia have been held most years since 1968, and the proceedings of these meetings have usually been published, although in a variety of forms.

The statutes of ICOHTEC were approved by the Constitutive Assembly of the IUHPS in Paris on 27 August 1968, and have subsequently been amended in 1974 and 1985. They consist of 23 Articles, setting out the rules for the government of the association. The aims are stated in Article 4:

(1) To establish close working relationships among specialists of different disciplines in order to foster international co-operation for the study and development of the history of technology.

(2) To promote the study of appropriate historical subjects, on the one hand by establishing and extending the scholarly

bases of the history of technology, and on the other by contributing to the resolution of certain contemporary national or international problems.

(3) To facilitate research and documentation for scholars in all countries in the history of technology by exchange of information and the creation of the material means necessary for this objective.

The rapid political changes which have transformed the world situation since 1991 have made it a matter of some urgency and importance to reform the Constitution of ICOHTEC so as to bring it fully into line with the new international realities.

The objective of this reform, agreed in Zaragoza at the XIXth International Congress of the History of Science and Technology, has been to put ICOHTEC on a basis of individual membership, with members paying an annual subscription of $25, but allowing also for institutional membership at a rate of $100, all such subscriptions to run from 1 January 1994. The following were elected to serve on the ICOHTEC Executive Committee:

President:	Professor Angus Buchanan
Vice Presidents:	Professor Carroll Pursell
	Professor Yuri Voronkov
Secretary-General:	Professor Hans-Joachim Braun
Treasurer:	Professor Alexandre Herlea
Journal Editor:	Dr Graham Hollister-Short

The subscription package will initially include a newsletter, appearing two or three times a year. It was felt that the annual newsletter does not provide members with a sufficient flow of up-to-date information and briefing on meetings, events and so forth, relating to the history of technology. It is planned as well to publish an ICOHTEC Yearbook. The yearbook will contain scholarly articles mainly based on papers presented at ICOHTEC symposia. Newsletters and Yearbook will all form part of the subscription package. In addition, a Programme Committee will seek to arrange a more extensive public programme of events (ICOHTEC Symposia) than has been the case previously.

So that ICOHTEC should be seen in reality, as well as in theory, to be a global organization for the history of technology, every effort will be made to achieve a world-wide membership base. This base is already strong in Europe and North America, but much work will have to be put in to extend membership in regions not so strongly represented up to now, notably South America and South East Asia. Colleagues wishing to become members should contact H.-J. Braun.

Apart from discussing topics in the history of technology, ICOHTEC's prime objective, ICOHTEC will also provide a forum for the discussion of present-day problems of technology and their relationship to socioeconomic, political, environmental and cultural issues.

ICOHTEC MEETINGS AND PUBLICATIONS

1.	Paris 1968:	XIIth International Congress Foundation of ICOHTEC: first symposium Papers published in Congress Proceedings.
2.	Pont-à-Mousson 1970:	ICOHTEC symposium: *L'acquisition des techniques par les pays non-initiateurs*, Paris 1973.
3.	Moscow 1971:	XIIIth International Congress ICOHTEC symposium Papers published in Congress Proceedings.
4.	Jablonna Poland 1973:	ICOHTEC symposium No publication.
5.	Tokyo 1974:	XIVth International Congress ICOHTEC symposium Papers published in Congress Proceedings.
6.	Kaluga USSR 1976:	ICOHTEC symposium: *Technology and Society* Published Prague 1976.
7.	Edinburgh 1977:	XVth International Congress ICOHTEC symposium: *The human implications of 20th century communications technology* No ICOHTEC publication.
8.	Stirling 1977:	Joint symposium with the Newcomen Society: *Scottish Engineers and Engineering* — Transactions of the Newcomen Society, Vol. 49, (1977–8), 121–48.
9.	Freiberg 1978:	ICOHTEC symposium: *Geschichte des Bergbaus und Hüttenwesens*, 2 vols, Freiberg 1978.
10.	Sofia 1980:	ICOHTEC symposium Local publication only.
11.	Bucharest 1981:	XVIth International Congress ICOHTEC symposium: *Technology, Humanism and Peace* Papers published in Congress Proceedings.
12.	Smolenice 1982:	ICOHTEC symposium: *Sources for the History of Technology — National Comparisons*, Prague 1983.
13.	Lerbach 1984:	ICOHTEC symposium: *Energie in der Geschichte*, 2 vols, Düsseldorf 1984.

14. Berkeley USA 1985:	XVIIth International Congress ICOHTEC symposium: *Technological Education — Technological Style*, San Francisco 1986.
15. Dresden 1986:	ICOHTEC symposium: *Teknik und Teknikwissenschaften in der Geschichte*, Berlin 1987.
16. Madrid 1988:	ICOHTEC symposium: *Civil Engineering between 1750 and 1850* Conference papers, CEHOPU, Madrid 1988.
17. Hamburg 1989:	XVIIIth International Congress ICOHTEC symposium: 'Failed Innovations' published in *Social Studies of Science*, Vol. 22, No. 2, May 1992.
18. Paris 1990:	ICOHTEC symposium: *Science-Technology Relationships*, San Francisco 1993.
19. Vienna 1991:	ICOHTEC symposium: *The Development of Technology in Traffic and Transport Systems*, Vienna 1993.
20. Uppsala 1992:	SHOT Annual Conference ICOHTEC symposium: *The Steam Engine as a Greek Temple:* *Art and Technology throughout History* Conference papers, Uppsala 1992.
21. Zaragoza 1993:	XIXth International Congress ICOHTEC symposium: *The Place of Theory in the History of Technology.*
22. Bath 1994:	ICOHTEC symposium: *International Themes in the History of Technology* (forthcoming)

Contents of Former Volumes

FIRST ANNUAL VOLUME, 1976*

D.S.L. CARDWELL and RICHARD L. HILLS, Thermodynamics and Practical Engineering in the Nineteenth Century.

JACQUES HEYMAN, Couplet's Engineering Memoirs, 1726–33.

NORMAN A.F. SMITH, Attitudes to Roman Engineering and the Question of the Inverted Siphon.

R.A. BUCHANAN, The Promethean Revolution: Science, Technology and History.

M. DAUMAS, The History of Technology: Its Aims, its Limits, its Methods.

KEITH DAWSON, Electromagnetic Telegraphy: Early Ideas, Proposals and Apparatus.

MARIE BOAS HALL, The Strange Case of Aluminium.

G. HOLLISTER-SHORT, Leads and Lags in Late Seventeenth-century English Technology.

SECOND ANNUAL VOLUME, 1977*

EMORY L. KEMP, Samuel Brown: Britain's Pioneer Suspension Bridge Builder.

DONALD R. HILL, The Banū Mūsà and their 'Book of Ingenious Devices'.

J.F. CAVE, A Note on Roman Metal Turning.

J.A. GARCIA-DIEGO, Old Dams in Extremadura.

G. HOLLISTER-SHORT, The Vocabulary of Technology.

RICHARD L. HILLS, Museums, History and Working Machines.

DENIS SMITH, The Use of Models in Nineteenth-century British Suspension Bridge Design.

NORMAN A.F. SMITH, The Origins of the Water Turbine and the Invention of its Name.

THIRD ANNUAL VOLUME, 1978

JACK SIMMONS, Technology in History.

R.A. BUCHANAN, History of Technology in the Teaching of History.

P.B. MORICE, The Role of History in a Civil Engineering Course.

JOYCE BROWN, Sir Proby Cautley (1802–71), a Pioneer of Indian Irrigation.

A. RUPERT HALL, On Knowing, and Knowing How to . . .

FRANK D. PRAGER, Vitruvius and the Elevated Aqueducts.

JAMES A. RUFFNER, Two Problems in Fuel Technology.

JOHN C. SCOTT, The Historical Development of Theories of Wave-Calming Oil.

FOURTH ANNUAL VOLUME, 1979

P.S. BARDELL, Some Aspects of the History of Journal Bearings and their Lubrication.

K.R. FAIRCLOUGH, The Waltham Pound Lock.

ROBERT FRIEDEL, Parkesine and Celluloid: The Failure and Success of the First Modern Plastic.

J.G. JAMES, Iron Arched Bridge Designs in Pre-Revolutionary France.

L.J. JONES, The Early History of Mechanical Harvesting.

G. HOLLISTER-SHORT, The Sector and Chain: An Historical Enquiry.

FIFTH ANNUAL VOLUME, 1980

THOMAS P. HUGHES, The Order of the Technological World.

THORKILD SCHIØLER, Bronze Roman Pistol Pumps.

STILLMAN DRAKE, Measurement in Galileo's Science.

L.J. JONES, John Ridley and the South Australian 'Stripper'.

D.G. TUCKER, Emile Lamm's Self-Propelled Tramcars 1870–72 and the Evolution of the Fireless Locomotive.

S.R. BROADBRIDGE, British Industry in 1767: Extracts from a Travel Journey of Joseph Banks.

RICHARD L. HILLS, Water, Stampers and Paper in the Auvergne: A Medieval Tradition.

SIXTH ANNUAL VOLUME, 1981*

MARJORIE NICE BOYER, Moving Ahead with the Fifteenth Century: New Ideas in Bridge Construction at Orleans.

ANDRÉ WEGENER SLEESWYK, Hand-Cranking in Egyptian Antiquity.

CHARLES SÜSSKIND, The Invention of Computed Tomography.

RICHARD L. HILLS, Early Locomotive Building near Manchester.

L.L. COATSWORTH, B.I. KRONBERG and M.C. USSELMAN, The Artefact as Historical Document. Part 1: The Fine Platinum Wires of W.H. Wollaston.

A. RUPERT HALL and N.C. RUSSELL, What about the Fulling-Mill?

MICHAEL FORES, *Technik*: Or Mumford Reconsidered.

SEVENTH ANNUAL VOLUME, 1982*

MARJORIE NICE BOYER, Water Mills: A Problem for the Bridges and Boats of Medieval France.

Wm. DAVID COMPTON, Internal-combustion Engines and their Fuel: A Preliminary Exploration of Technological Interplay.

F.T. EVANS, Wood since the Industrial Revolution: A Strategic Retreat?

MICHAEL FORES, Francis Bacon and the Myth of Industrial Science.

D.G. TUCKER, The Purpose and Principles of Research in an Electrical Manufacturing Business of Moderate Size, as Stated by J.A. Crabtree in 1930.

ROMAN MALINOWSKI, Ancient Mortars and Concretes: Aspects of their Durability.

V. FOLEY, W. SOEDEL, J. TURNER and B. WILHOITE, The Origin of Gearing.

EIGHTH ANNUAL VOLUME, 1983*

W. ADDIS, A New Approach to the History of Structural Engineering.

HANS-JOACHIM BRAUN, The National Association of German-American Technologists and Technology Transfer between Germany and the United States, 1884–1930.

W. BERNARD CARLSON, Edison in the Mountains: The Magnetic Ore Separation Venture, 1879–1900.

THOMAS DAY, Samuel Brown: His Influence on the Design of Suspension Bridges.

ROBERT H.J. SELLIN, The Large Roman Water Mill at Barbegal (France).

G. HOLLISTER-SHORT, The Use of Gunpowder in Mining: A Document of 1627.

MIKULÁŠ TEICH, Fermentation Theory and Practice: The Beginnings of Pure Yeast Cultivation and English Brewing, 1883–1913.

GEORGE TIMMONS, Education and Technology in the Industrial Revolution.

NINTH ANNUAL VOLUME, 1984*

P.S. BARDELL, The Origins of Alloy Steels.

MARJORIE NICE BOYER, A Fourteenth-Century Pile Driver: the *Engin* of the Bridge at Orléans.

MICHAEL DUFFY, Rail Stresses, Impact Loading and Steam Locomotive Design.

JOSÉ A. GARCIA-DIEGO, Giovanni Francesco Sitoni, an Hydraulic Engineer of the Renaissance.

DONALD R. HILL, Information on Engineering in the Works of Muslim Geographers.

ROBERT J. SPAIN, The Second-Century Romano-British Watermill at Ickham, Kent.

IAN R. WINSHIP, The Gas Engine in British Agriculture, *c.* 1870–1925.

TENTH ANNUAL VOLUME, 1985*

D de COGAN, Dr. E.O.W. Whitehouse and the 1858 trans-Atlantic Cable.

A. RUPERT HALL, Isaac Newton's Steamer.

G.J. HOLLISTER-SHORT, Gunpowder and Mining in Sixteenth- and Seventeenth-Century Europe.

G.J. JACKSON, Evidence of American Influence on the Designs of Nineteenth-Century Drilling Tools, Obtained from British Patent Specifications and Other Sources.

JACQUES PAYEN, Beau de Rochas Devant la Technique et l'Industrie de son Temps.

ÖRJAN WIKANDER, Archaeological Evidence for Early Water-Mills — An Interim Report.

A.P. WOOLRICH, John Farey and the Smeaton Manuscripts.

MIKE CHRIMES, Bridges: A Bibliography of Articles Published in Scientific Periodicals 1800–1829.

ELEVENTH ANNUAL VOLUME, 1986*

HANS-JOACHIM BRAUN, Technology Transfer under Conditions of War: German Aero-technology in Japan during the Second World War.

VERNARD FOLEY, with SUSAN CANGANELLI, JOHN CONNOR and DAVID RADER, Using the Early Slide-rest.

J.G. JAMES, The Origins and Worldwide Spread of Warren-truss Bridges in the Mid-nineteenth Century. Part 1: Origins and Early Examples in the UK.

ANDREW NAHUM, The Rotary Aero Engine.

DALE. H. PORTER, An Historian's Judgments about the Thames Embankment.

JOHN H. WHITE, More Than an Idea Whose Time Has Come: The Beginnings of Steel Freight Cars.

IAN R. WINSHIP, The Acceptance of Continuous Brakes on Railways in Britain.

TWELFTH ANNUAL VOLUME, 1990

KENNETH C. BARRACLOUGH, Swedish Iron and Sheffield Steel.

IAN INKSTER, Intellectual Dependency and the Sources of Invention: Britain and the Australian Technological System in the Nineteenth Century.

M.T. WRIGHT, Rational and Irrational Reconstruction: The London Sundial-Calendar and the Early History of Geared Mechanisms.

J.V. FIELD, Some Roman and Byzantine Portable Sundials and the London Sundial-Calendar.

R.T. McCUTCHEON, Modern Construction Technology in Low-income Housing Policy: The Case of Industrialized Building and the Manifold Links between Technology and Society in an Established Industry.

THIRTEENTH ANNUAL VOLUME, 1991

FOURTEENTH ANNUAL VOLUME, 1992

JOHN LANGDON, The Birth and Demise of a Medieval Windmill

MICHAEL J.T. LEWIS, The South-pointing Chariot in Rome: Gearing in China and the West.

STANLEY SMITH, The Development and Use of the Tubular Beam, 1830–1860.

NICOLÁS GARCÍA TAPIA, Some Designs of Jerónimo de Ayanz (*c.* 1550–1613) relating to Mining, Metallurgy and Steam Pumps.

DONALD E. TARTER, Peenemünde and Los Alamos: Two Studies.

GRAHAM HOLLISTER-SHORT, Reflections on Two Conferences: ICOHTEC XIX and MASTECH.

Announcement: The Georgius Agricola Commemorations, 1994.

FIFTEENTH ANNUAL VOLUME, 1993

MICHAEL E. GORMAN, MATTHEW M. MEHALIK, W. BERNARD CARLSON AND MICHAEL OBLON, Alexander Graham Bell, Elisha Gray and the Speaking Telegraph: A Cognitive Comparison.

GRAHAM HOLLISTER-SHORT, On the Origins of the Suction Lift Pump.

ALEX KELLER, Technological Aspirations and the Motivation of Natural Philosophy in Seventeenth-Century England.

WALTER ENDREI, Count Theodore Batthyány's Paddle-Wheel Ship.

BJÖRN IVAR BERG, The Kongsberg Silver Mines and the Norwegian Mining Museum.

HERMANN KNOFLACHER, Does the Development of Mobility in Traffic Follow a Pattern?

MICHAEL J.T. LEWIS, The Greeks and the Early Windmill.

GRAEME GOODAYE, Faraday Reinvented: Moral Imagery and Institutional Icons in Victorian Electrical Engineering.

PHILIPPE BRAUNSTEIN, Légendes Welsches et Itinéraires Silésiens: La Prospection Minière au XVe Siècle (English summary by G.H-S).

* Out of print